Solutions Guide

Introductory Chemistry: A Foundation
Introductory Chemistry
Basic Chemistry

Fourth Edition

Zumdahl

James F. Hall
University of Massachusetts Lowell

Houghton Mifflin Company
Boston   New York

Editor-in-Chief: Kathi Prancan
Associate Editor: Marianne Stepanian
Editorial Assistants: Sarah Gessner and Joy Park
Senior Manufacturing Coordinator: Marie Barnes
Executive Marketing Manager: Andy Fisher

Copyright © 2000 by Houghton Mifflin Company. All rights reserved.

No part of this work may be reproduced or transmitted in any form or by any means, electronic or mechanical, including photocopying and recording, or by any information storage or retrieval system without the prior written permission of Houghton Mifflin Company unless such copying is expressly permitted by federal copyright law. Address inquiries to College Permissions, Houghton Mifflin Company, 222 Berkeley Street, Boston, MA 02116-3764.

Printed in the U.S.A.

ISBN: 0-395-95543-2

3 4 5 6 7 8 9-CS-03 02 01 00

Table of Contents

Preface..iv
Chapter 1   Chemistry: An Introduction..................................1
Chapter 2   Measurements and Calculations...............................3
Chapter 3   Matter and Energy..16
Cumulative Review: Chapters 1, 2, and 3................................23
Chapter 4   Chemical Foundations: Elements, Atoms, and Ions...........28
Chapter 5   Nomenclature...37
Cumulative Review: Chapters 4 and 5....................................45
Chapter 6   Chemical Reactions: An Introduction........................50
Chapter 7   Reactions in Aqueous Solutions.............................55
Cumulative Review: Chapters 6 and 7....................................66
Chapter 8   Chemical Composition.......................................70
Chapter 9   Chemical Quantities..98
Cumulative Review: Chapters 8 and 9...................................119
Chapter 10   Modern Atomic Theory......................................126
Chapter 11   Chemical Bonding..133
Cumulative Review: Chapters 10 and 11.................................147
Chapter 12   Gases...153
Chapter 13   Liquids and Solids..170
Chapter 14   Solutions...176
Cumulative Review: Chapters 12, 13, and 14............................193
Chapter 15   Acids and Bases...200
Chapter 16   Equilibrium...210
Cumulative Review: Chapters 15 and 16..................................219
Chapter 17   Oxidation-Reduction Reactions/Electrochemistry............226
Chapter 18   Radioactivity and Nuclear Energy..........................239
Chapter 19   Organic Chemistry...245
Chapter 20   Biochemistry..258

Preface

This guide contains detailed solutions for the even-numbered end-of-chapter problems in the fourth editions of *Introductory Chemistry*, *Introductory Chemistry: A Foundation*, and *Basic Chemistry* by Steven S. Zumdahl. Hundreds of new problems and questions have been prepared for the fourth editions of the text, which we hope will be of help to you in gaining an understanding of the fundamental principles of chemistry.

Think of this guide as the last chapter of a mystery novel. Just as you wouldn't initially turn to the last few pages of a mystery to find out "whodunit," you should not look at the solutions in this guide until you have tried diligently to solve the end-of-chapter problems yourself. To learn chemistry effectively, you have to spend a lot of time pushing a pencil around on paper!

As you study chemistry, first go over your lecture notes and read through your textbook to see if there is any factual material or terminology you don't understand. Then work through the example problems your professor had done in class and the example problems and self-check exercises in the textbook, paying close attention to the *method* of solution used for each problem. Then try again to work through the *same* classroom and textbook examples yourself *without* looking at the solutions. Once you feel comfortable with the example problems, you can feel confident to tackle the end-of-chapter problems. After you've mastered a few chapters, then test your understanding by going through the *Cumulative Review* material provided for those chapters. These reviews tie together the material covered in the previous few chapters and are intended to give you some perspective on your progress in chemistry.

One topic that causes students much concern is the matter of significant figures, and the determination of the number of digits to which a solution to a problem should be reported. To avoid truncation errors in the solutions contained in this guide, the solutions typically report *intermediate* answers to one more digit that appropriate for the final answer. The *final* answer to each problem is then given to the correct number of significant figures based on the data provided in the problem. Realize especially that your calculator doesn't know anything about the rules for significant figures: a calculator will report answers to however many digits it is programmed for, regardless of whether those digits mean anything. *You* have to have the understanding of the measurements involved and have to do the rounding of answers yourself.

Many very dedicated people have worked long hours to prepare this guide. Particular thanks go to Leslie di Verdi of Colorado State University for carefully checking the manuscript for accuracy, and to Sarah Gessner, Editorial Assistant at Houghton Mifflin, for her patience, competence, and good cheer.

Good luck in your study of chemistry!

James F. Hall
James_Hall@uml.edu

Chapter 1 Chemistry: An Introduction

2. Examples: physician (understanding cellular processes, understanding how drugs and bloodtests work); lawyer (understanding scientific/forensic laboratory tests for use in court); pharmacist (understanding how drugs work, and interactions between drugs); artist (understanding the various media used in art work); photographer (understanding how the film exposure and developing chemical processes occur and how they can be controlled and modified); farmer (understanding which pesticide or fertilizer is needed and how these chemicals work); nurse (understanding how various tests and drugs may affect the patient's wellbeing).

4. There are, unfortunately, many examples. Chemical and biological weapons are still being produced in some countries. Although the development of new plastics has been a boon in many endeavors, this also increases our depletion of fossil fuels and our solid waste problems. Although many exciting new drugs and treatments have become available, the same biotechnology may lead to testing procedures for determining whether a person has a genetic likelihood of developing a particular disease, which may make it impossible or difficult for that person to obtain health or life insurance.

6. This answer depends on your own experience, but consider the following examples: oven cleaner (the label says it contains sodium hydroxide; it converts the burned on grease in the oven to a soapy material that washes away); drain cleaner (the label says it contains sodium hydroxide; it dissolves the clog of hair in the drain); stomach antacid (the label says it contains calcium carbonate; it makes me belch and makes my stomach feel better); hydrogen peroxide (the label says it is a 3% solution of hydrogen peroxide; when applied to a wound, it bubbles); depilatory cream (the label says it contains sodium hydroxide; it removes unwanted hair from skin).

8. Answer will depend on student experience.

10.
    a. quantitative - a number (measurement) is indicated explicitly
    b. qualitative - only a qualitative description is given
    c. quantitative - a numerical measurement is indicated
    d. qualitative - only a qualitative description is given
    e. quantitative - a number (measurement) is implied
    f. qualitative - a qualitative judgment is given
    g. quantitative - a numerical quantity is indicated

12. A natural law is a *summary of observed, measurable behavior* that occurs repeatedly and consistently. A theory is our attempt to *explain* such behavior. The conservation of mass observed during chemical reactions is an example of a natural law. The idea that the universe began with a "big bang" is an example of a theory.

Chapter 1   Chemistry: An Introduction

14. Most applications of chemistry are oriented toward the interpretation of observations and the solving of problems. Although memorization of some facts may *aid* in these endeavors, it is the ability to combine, relate, and synthesize information that is most important in the study of chemistry.

16. In real life situations, the problems and applications likely to be encountered are not simple textbook examples. One must be able to observe an event, hypothesize a cause, and then test this hypothesis. One must be able to carry what has been learned in class forward to new, different situations.

Chapter 2   Measurements and Calculations

2. 10^4

4. Because 0.0021 is less than one, the exponent will be *negative*. Because 4540 is greater than one, the exponent will be *positive*.

6. a. -5; 6.7×10^{-5}

   b. 6; 9.331442×10^6

   c. -4; 1×10^{-4}

   d. 4; 1.631×10^4

8. a. The decimal point must be moved six places to the left, so the exponent is positive 6; $9{,}367{,}421 = 9.367421 \times 10^6$

   b. The decimal point must be moved three places to the left, so the exponent is positive 3; $7241 = 7.241 \times 10^3$

   c. The decimal point must be moved four places to the right, so the exponent is negative 4; $0.0005519 = 5.519 \times 10^{-4}$

   d. The decimal point does not have to be moved, so the exponent is zero; $5.408 = 5.408 \times 10^0$

   e. 6.24×10^2 is already written in standard scientific notation.

   f. The decimal point must be moved three places to the left, and the resulting exponent of positive three must be combined with the exponent of negative two in the multiplier; $6{,}319 \times 10^{-2} = 6.319 \times 10^1$

   g. The decimal point must be moved nine places to the right, so the exponent is negative nine; $0.000000007215 = 7.215 \times 10^{-9}$

   h. The decimal point must be moved one place to the right, so the exponent is negative 1; $0.721 = 7.21 \times 10^{-1}$

10. a. The decimal point must be moved two places to the right; $4.83 \times 10^2 = 483$

    b. The decimal point must be moved four places to the left; $7.221 \times 10^{-4} = 0.0007221$

    c. The decimal point does not have to be moved; $6.1 \times 10^0 = 6.1$

    d. The decimal point must be moved eight places to the left; $9.11 \times 10^{-8} = 0.0000000911$

    e. The decimal point must be moved six places to the right; $4.221 \times 10^6 = 4{,}221{,}000$

f. The decimal point must be moved three places to the left; $1.22 \times 10^{-3} = 0.00122$

g. The decimal point must be moved three places to the right; $9.999 \times 10^3 = 9,999$

h. The decimal point must be moved five places to the left; $1.016 \times 10^{-5} = 0.00001016$

i. The decimal point must be moved five places to the right; $1.016 \times 10^5 = 101,600$

j. The decimal point must be moved one place to the left; $4.11 \times 10^{-1} = 0.411$

k. The decimal point must be moved four places to the right; $9.71 \times 10^4 = 97,100$

l. The decimal point must be moved four places to the left; $9.71 \times 10^{-4} = 0.000971$

12. To say that scientific notation is in *standard* form means that you have a number between 1 and 10, followed by an exponential term. The numbers given in this problem are *not* between 1 and 10 as written.

    a. $142.3 \times 10^3 = (1.423 \times 10^2) \times 10^3 = 1.423 \times 10^5$

    b. $0.0007741 \times 10^{-9} = (7.741 \times 10^{-4}) \times 10^{-9} = 7.741 \times 10^{-13}$

    c. $22.7 \times 10^3 = (2.27 \times 10^1) \times 10^3 = 2.27 \times 10^4$

    d. 6.272×10^{-5} is already written in standard scientific notation.

    e. $0.0251 \times 10^4 = (2.51 \times 10^{-2}) \times 10^4 = 2.51 \times 10^2$

    f. $97,522 \times 10^{-3} = (9.7522 \times 10^4) \times 10^{-3} = 9.7522 \times 10^1$

    g. $0.0000097752 \times 10^6 = (9.7752 \times 10^{-6}) \times 10^6 = 9.7752 \times 10^0$ (9.97752)

    h. $44,252 \times 10^4 = (4.4252 \times 10^4) \times 10^4 = 4.4252 \times 10^8$

14. a. $1/0.00032 = 3.1 \times 10^3$

    b. $10^3/10^{-3} = 1 \times 10^6$

    c. $10^3/10^3 = 1$ (1×10^0); any number divided by itself is unity.

    d. $1/55,000 = 1.8 \times 10^{-5}$

    e. $(10^5)(10^4)(10^{-4})/10^{-2} = 1 \times 10^7$

    f. $43.2/(4.32 \times 10^{-5}) = \dfrac{4.32 \times 10^1}{4.32 \times 10^{-5}} = 1.00 \times 10^6$

Chapter 2   Measurements and Calculations   5

      g. $(4.32 \times 10^{-5})/432 = \dfrac{4.32 \times 10^{-5}}{4.32 \times 10^2} = 1.00 \times 10^{-7}$

      h. $1/(10^5)(10^{-6}) = 1/(10^{-1}) = 1 \times 10^1$

16. grams

18.
   a. mega-
   b. milli-
   c. nano-
   d. mega-
   e. centi-
   f. micro-

20. A mile represents, by definition, a greater distance than a kilometer. Therefore 100 mi represents a greater distance than 100 km.

22. quart

24. kilogram

26. 1.62 m is approximately 5 ft, 4". The woman is slightly taller.

28. d

30. d (the other units would give very large numbers for the distance)

32. Table 2.6 indicates that the diameter of a quarter is 2.5 cm.

     $1 \text{ m} \times \dfrac{100 \text{ cm}}{1 \text{ m}} \times \dfrac{1 \text{ quarter}}{2.5 \text{ cm}} = 40 \text{ quarters}$

34. uncertainty

36. The scale of the ruler shown is only marked to the nearest *tenth* of a centimeter; writing 2.850 would imply that the scale was marked to the nearest *hundredth* of a centimeter (and that the zero in the thousandths place had been estimated).

38.
   a. probably only two
   b. infinite (a definition)
   c. infinite (a definition)
   d. probably only 1
   e. three (the race is defined to be exactly 500. miles)

40. final

Chapter 2   Measurements and Calculations

42. a. 4.23×10^{-1}
     b. 7.12×10^{6}
     c. 4.45×10^{-4}
     d. 2.30×10^{-4}
     e. 9.72×10^{5}

44. a. 3.42×10^{-4}
     b. 1.034×10^{4}
     c. 1.7992×10^{1}
     d. 3.37×10^{5}

46. decimal

48. three

50. none

52. a. 641.0 (the answer can only be given to one decimal place, since 212.7 and 26.7 are only given to one decimal place)

     b. 1.327 (the answer can only be given to three decimal places, since 0.221 is only given to three decimal places)

     c. 77.34 (the answer can only be given to two decimal places, since 26.01 is only given to two decimal places)

     d. Before performing the calculation, the numbers have to be converted so that they contain the same power of ten.
         $2.01 \times 10^{2} + 3.014 \times 10^{3} = 2.01 \times 10^{2} + 30.14 \times 10^{2} = 32.15 \times 10^{2}$
         This answer should then be converted to *standard* scientific notation, $32.15 \times 10^{2} = 3.215 \times 10^{3} = 3,215$.

54. a. 124 (the answer can only be given to three significant figures because 0.995 is only given to three significant figures)

     b. 1.995×10^{-23} (the answer can only be given to four significant figures because 6.022×10^{23} is only given to four significant figures)

     c. 1.14×10^{-2} (the answer can only be given to three significant figures because 0.500 is only given to three significant figures.)

     d. 5.3×10^{-4} (the answer can only be given to two significant figures because 0.15 is only given to two significant figures)

56. a. $(2.0944 + 0.0003233 + 12.22)/7.001 =$
         $(14.3147233)/7.001 = 2.045$

     b. $(1.42 \times 10^{2} + 1.021 \times 10^{3})/(3.1 \times 10^{-1}) =$
         $(142 + 1021)/(3.1 \times 10^{-1}) = (1163)/(3.1 \times 10^{-1}) = 3751 = 3.8 \times 10^{3}$

Chapter 2   Measurements and Calculations   7

c. $(9.762 \times 10^{-3})/(1.43 \times 10^{2} + 4.51 \times 10^{1}) =$
$(9.762 \times 10^{-3})/(143 + 45.1) = (9.762 \times 10^{-3})/(188.1) = 5.19 \times 10^{-5}$

d. $(6.1982 \times 10^{-4})^{2} = (6.1982 \times 10^{-4})(6.1982 \times 10^{-4}) = 3.8418 \times 10^{-7}$

58. an infinite number (a definition)

60. $\dfrac{1000 \text{ mL}}{1 \text{ L}} \triangleleft \quad \dfrac{1 \text{ L}}{1000 \text{ mL}}$

62. $\dfrac{1 \text{ lb}}{\$0.79}$

64. a. $2.23 \text{ m} \times \dfrac{1.094 \text{ yd}}{1 \text{ m}} = 2.44 \text{ yd}$

b. $46.2 \text{ yd} \times \dfrac{1 \text{ m}}{1.094 \text{ yd}} = 42.2 \text{ m}$

c. $292 \text{ cm} \times \dfrac{1 \text{ in.}}{2.54 \text{ cm}} = 115 \text{ in.}$

d. $881.2 \text{ in.} \times \dfrac{2.54 \text{ cm}}{1 \text{ in.}} = 2238 \text{ cm}$

e. $1043 \text{ km} \times \dfrac{1 \text{ mi}}{1.6093 \text{ km}} = 648.1 \text{ mi}$

f. $445.5 \text{ mi} \times \dfrac{1.6093 \text{ km}}{1 \text{ mi}} = 716.9 \text{ km}$

g. $36.2 \text{ m} \times \dfrac{1 \text{ km}}{1000 \text{ m}} = 0.0362 \text{ km}$

h. $0.501 \text{ km} \times \dfrac{1000 \text{ m}}{1 \text{ km}} \times \dfrac{100 \text{ cm}}{1 \text{ m}} = 5.01 \times 10^{4} \text{ cm}$

66. a. $254.3 \text{ g} \times \dfrac{1 \text{ kg}}{1000 \text{ g}} = 0.2543 \text{ kg}$

b. $2.75 \text{ kg} \times \dfrac{1000 \text{ g}}{1 \text{ kg}} = 2.75 \times 10^{3} \text{ g}$

c. $2.75 \text{ kg} \times \dfrac{1 \text{ lb}}{0.45359 \text{ kg}} = 6.06 \text{ lb}$

d. $2.75 \text{ kg} \times \dfrac{1 \text{ lb}}{0.45359 \text{ kg}} \times \dfrac{16 \text{ oz}}{1 \text{ lb}} = 97.0 \text{ oz}$

e. $534.1 \text{ g} \times \dfrac{1 \text{ lb}}{453.59 \text{ g}} = 1.177 \text{ lb}$

8   Chapter 2   Measurements and Calculations

    f. $1.75 \text{ lb} \times \dfrac{453.59 \text{ g}}{1 \text{ lb}} = 794 \text{ g}$

    g. $8.7 \text{ oz} \times \dfrac{1 \text{ lb}}{16 \text{ oz}} \times \dfrac{453.59 \text{ g}}{1 \text{ lb}} = 2.5 \times 10^2 \text{ g}$

    h. $45.9 \text{ g} \times \dfrac{1 \text{ lb}}{453.59 \text{ g}} \times \dfrac{16 \text{ oz}}{1 \text{ lb}} = 1.62 \text{ oz}$

68.   190 mi = 1.9×10^2 mi to two significant figures

    $1.9 \times 10^2 \text{ mi} \times \dfrac{1 \text{ km}}{0.62137 \text{ mi}} = 3.1 \times 10^2 \text{ km}$

    $3.1 \times 10^2 \text{ km} \times \dfrac{1000 \text{ m}}{1 \text{ km}} = 3.1 \times 10^5 \text{ m}$

    $1.9 \times 10^2 \text{ mi} \times \dfrac{5{,}280 \text{ ft}}{1 \text{ mi}} = 1.0 \times 10^6 \text{ ft}$

70. $1 \times 10^{-10} \text{ m} \times \dfrac{100 \text{ cm}}{1 \text{ m}} = 1 \times 10^{-8} \text{ cm}$

    $1 \times 10^{-8} \text{ cm} \times \dfrac{1 \text{ in.}}{2.54 \text{ cm}} = 4 \times 10^{-9} \text{ in.}$

    $1 \times 10^{-8} \text{ cm} \times \dfrac{1 \text{ m}}{100 \text{ cm}} \times \dfrac{10^9 \text{ nm}}{1 \text{ m}} = 0.1 \text{ nm}$

72.   Celsius

74.   273

76.   Fahrenheit (F)

78. $t_C = t_K - 273$

    a.   275 − 273 = 2°C

    b.   445 − 273 = 172°C

    c.   0 − 273 = −273°C

    d.   77 − 273 = −196°C

    e.   10,000. − 273 = 9727°C

    f.   2 − 273 = −271°C

80. $t_F = 1.80(t_C) + 32$

    a.   1.80(78.1) + 32 = 173 °F

b.   1.80(40.) + 32 = 104 °F

c.   1.80(-273) + 32 = -459 °F

d.   1.80(32) + 32 = 90. °F

82. a. $t_C = (t_F - 32)/1.80 = (-201 \text{ °F} - 32)/1.80 = (-233)/1.80 = -129.4$ °C

-129.4 °C + 273 = 143.6 = 144 K

b.   -201 °C + 273 = 72 K

c. $t_F = 1.80(t_C) + 32 = 1.80(351 \text{ °C}) + 32 = 664$ °F

d. $t_C = (t_F - 32)/1.80 = (-150 \text{ °F} - 32)/1.80 = -101$ °C

84. g/cm^3 (g/mL)

86. 100 in^3

88. Density is a *characteristic* property of a pure substance; all samples of the same pure substance have the *same* density.

90. copper

92. density = $\dfrac{\text{mass}}{\text{volume}}$

a. $d = \dfrac{122.4 \text{ g}}{5.5 \text{ cm}^3} = 22$ g/cm^3

b. $v = 0.57 \text{ m}^3 \times \left(\dfrac{100 \text{ cm}}{1 \text{ m}}\right)^3 = 5.7 \times 10^5$ cm^3

$d = \dfrac{1.9302 \times 10^4 \text{ g}}{5.7 \times 10^5 \text{ cm}^3} = 0.034$ g/cm^3

c. $m = 0.0175 \text{ kg} \times \dfrac{1000 \text{ g}}{1 \text{ kg}} = 17.5$ g

$d = \dfrac{17.5 \text{ g}}{18.2 \text{ mL}} = 0.962$ g/mL = 0.962 g/cm^3

d. $v = 0.12 \text{ m}^3 \times \left(\dfrac{100 \text{ cm}}{1 \text{ m}}\right)^3 = 1.2 \times 10^5$ cm^3

$d = \dfrac{2.49 \text{ g}}{1.2 \times 10^5 \text{ cm}^3} = 2.1 \times 10^{-5}$ g/cm^3

94. $d = \dfrac{75.2 \text{ g}}{89.2 \text{ mL}} = 0.843$ g/mL

96. $m = 3.5 \text{ lb} \times \dfrac{453.59 \text{ g}}{1 \text{ lb}} = 1.59 \times 10^3 \text{ g}$

$v = 1.2 \times 10^4 \text{ in}^3 \times \left(\dfrac{2.54 \text{ cm}}{1 \text{ in}}\right)^3 = 1.97 \times 10^5 \text{ cm}^3$

$d = \dfrac{1.59 \times 10^3 \text{ g}}{1.97 \times 10^5 \text{ cm}^3} = 8.1 \times 10^{-3} \text{ g/cm}^3$

The material will float.

98. $5.25 \text{ g} \times \dfrac{1 \text{ cm}^3}{10.5 \text{ g}} = 0.500 \text{ cm}^3 = 0.500 \text{ mL}$

$11.2 \text{ mL} + 0.500 \text{ mL} = 11.7 \text{ mL}$

100. a. $50.0 \text{ cm}^3 \times \dfrac{19.32 \text{ g}}{1 \text{ cm}^3} = 966 \text{ g}$

b. $50.0 \text{ cm}^3 \times \dfrac{7.87 \text{ g}}{1 \text{ cm}^3} = 394 \text{ g}$

c. $50.0 \text{ cm}^3 \times \dfrac{11.34 \text{ g}}{1 \text{ cm}^3} = 567 \text{ g}$

d. $50.0 \text{ cm}^3 \times \dfrac{2.70 \text{ g}}{1 \text{ cm}^3} = 135 \text{ g}$

102. a. $3.011 \times 10^{23} = 301,100,000,000,000,000,000,000$
b. $5.091 \times 10^9 = 5,091,000,000$
c. $7.2 \times 10^2 = 720$
d. $1.234 \times 10^5 = 123,400$
e. $4.32002 \times 10^{-4} = 0.000432002$
f. $3.001 \times 10^{-2} = 0.03001$
g. $2.9901 \times 10^{-7} = 0.00000029901$
h. $4.2 \times 10^{-1} = 0.42$

104. a. centimeters
b. meters
c. kilometers
d. centimeters
e. millimeters

106. a. $36.2 \text{ blim} \times \dfrac{1400 \text{ kryll}}{1 \text{ blim}} = 5.07 \times 10^4 \text{ kryll}$

b. $170 \text{ kryll} \times \dfrac{1 \text{ blim}}{1400 \text{ kryll}} = 0.12 \text{ blim}$

Chapter 2 Measurements and Calculations 11

c. $72.5 \text{ kryll}^2 \times \left(\dfrac{1 \text{ blim}}{1400 \text{ kryll}}\right)^2 = 3.70 \times 10^{-5} \text{ blim}^2$

108. $52 \text{ cm} \times \dfrac{1 \text{ in.}}{2.54 \text{ cm}} = 20. \text{ in.}$

110. $1 \text{ lb} \times \dfrac{1 \text{ kg}}{2.2 \text{ lb}} \times \dfrac{\$1}{F5} \times \dfrac{11.5F}{1 \text{ kg}} = \$1$

112. °X = 1.26°C + 14

114. $d = \dfrac{36.8 \text{ g}}{10.5 \text{ L}} = 3.50 \text{ g/L} \quad (3.50 \times 10^{-3} \text{ g/cm}^3)$

116. for ethanol, $100. \text{ mL} \times \dfrac{0.785 \text{ g}}{1 \text{ mL}} = 78.5 \text{ g}$

for benzene, $1000 \text{ mL} \times \dfrac{0.880 \text{ g}}{1 \text{ mL}} = 880. \text{ g}$

total mass, 78.5 + 880. = 959 g

118. a. negative
 b. negative
 c. positive
 d. zero
 e. negative

120. a. 2; positive
 b. 11; negative
 c. 3; positive
 d. 5; negative
 e. 5; positive
 f. 0; zero
 g. 1; negative
 h. 7; negative

122. a. 1; positive
 b. 3; negative
 c. 0; zero
 d. 3; positive
 e. 9; negative

Chapter 2   Measurements and Calculations

124. a. The decimal point must be moved five places to the left; $2.98 \times 10^{-5} = 0.0000298$

b. The decimal point must be moved nine places to the right; $4.358 \times 10^9 = 4{,}358{,}000{,}000$

c. The decimal point must be moved six places to the left; $1.9928 \times 10^{-6} = 0.0000019928$

d. The decimal point must be moved 23 places to the right; $6.02 \times 10^{23} = 602{,}000{,}000{,}000{,}000{,}000{,}000{,}000$

e. The decimal point must be moved one place to the left; $1.01 \times 10^{-1} = 0.101$

f. The decimal point must be moved three places to the left; $7.87 \times 10^{-3} = 0.00787$

g. The decimal point must be moved seven places to the right; $9.87 \times 10^7 = 98{,}700{,}000$

h. The decimal point must be moved two places to the right; $3.7899 \times 10^2 = 378.99$

i. The decimal point must be moved one place to the left; $1.093 \times 10^{-1} = 0.1093$

j. The decimal point must be moved zero places; $2.9004 \times 10^0 = 2.9004$

k. The decimal point must be moved four places to the left; $3.9 \times 10^{-4} = 0.00039$

l. The decimal point must be moved eight places to the left; $1.904 \times 10^{-8} = 0.00000001904$

126. a. $1/10^2 = 1 \times 10^{-2}$

b. $1/10^{-2} = 1 \times 10^2$

c. $55/10^3 = \dfrac{5.5 \times 10^1}{1 \times 10^3} = 5.5 \times 10^{-2}$

d. $(3.1 \times 10^6)/10^{-3} = \dfrac{3.1 \times 10^6}{1 \times 10^{-3}} = 3.1 \times 10^9$

e. $(10^6)^{1/2} = 1 \times 10^3$

f. $(10^6)(10^4)/(10^2) = \dfrac{(1 \times 10^6)(1 \times 10^4)}{(1 \times 10^2)} = 1 \times 10^8$

g. $1/0.0034 = \dfrac{1}{3.4 \times 10^{-3}} = 2.9 \times 10^2$

Chapter 2   Measurements and Calculations   13

      h. $3.453/10^{-4} = \dfrac{3.453}{1 \times 10^{-4}} = 3.453 \times 10^{4}$

128. kelvin, K

130. centimeter

132. 0.105 m

134. 1 kg (100 g = 0.1 kg)

136. 10 cm (1 cm = 10 mm)

138. 2.8 (the hundredths place is estimated)

140.
- a. 0.000426
- b. 4.02×10^{-5}
- c. 5.99×10^{6}
- d. 400.
- e. 0.00600

142.
- a. 2149.6 (the answer can only be given to the first decimal place, since 149.2 is only known to the first decimal place)

- b. 5.37×10^{3} (the answer can only be given to two decimal places since 4.34 is only known to two decimal places; since the power of ten is the same for each number, the calculation can be performed directly)

- c. Before performing the calculation, the numbers have to be converted so that they contain the same power of ten.

    $4.03 \times 10^{-2} - 2.044 \times 10^{-3} =$

    $4.03 \times 10^{-2} - 0.2044 \times 10^{-2} =$

    3.83×10^{-2} (the answer can only be given the second decimal place since 4.03×10^{-2} is only known to the second decimal place)

- d. Before performing the calculation, the numbers have to be converted so that they contain the same power of ten.

    $2.094 \times 10^{5} - 1.073 \times 10^{6} =$

    $2.094 \times 10^{5} - 10.73 \times 10^{5} =$

    -8.64×10^{5}

14   Chapter 2   Measurements and Calculations

144. a. $(2.9932 \times 10^4)(2.4443 \times 10^2 + 1.0032 \times 10^1) =$

    $(2.9932 \times 10^4)(24.443 \times 10^1 + 1.0032 \times 10^1) =$

    $(2.9932 \times 10^4)(25.446 \times 10^1) = 7.6166 \times 10^6$

  b. $(2.34 \times 10^2 + 2.443 \times 10^{-1})/(0.0323) =$

    $(2.34 \times 10^2 + 0.002443 \times 10^2)/(0.0323) =$

    $(2.34 \times 10^2)/(0.0323) = 7.24 \times 10^3$

  c. $(4.38 \times 10^{-3})^2 = 1.92 \times 10^{-5}$

  d. $(5.9938 \times 10^{-6})^{1/2} = 2.4482 \times 10^{-3}$

146. 1 year/12 months; 12 months/1 year

148. a. $908 \text{ oz} \times \dfrac{1 \text{ lb}}{16 \text{ oz}} \times \dfrac{1 \text{ kg}}{2.2046 \text{ lb}} = 25.7 \text{ kg}$

  b. $12.8 \text{ L} \times \dfrac{1 \text{ qt}}{0.94633 \text{ L}} \times \dfrac{1 \text{ gal}}{4 \text{ qt}} = 3.38 \text{ gal}$

  c. $125 \text{ mL} \times \dfrac{1 \text{ L}}{1000 \text{ mL}} \times \dfrac{1 \text{ qt}}{0.94633 \text{ L}} = 0.132 \text{ qt}$

  d. $2.89 \text{ gal} \times \dfrac{4 \text{ qt}}{1 \text{ gal}} \times \dfrac{1 \text{ L}}{1.0567 \text{ qt}} \times \dfrac{1000 \text{ mL}}{1 \text{ L}} = 1.09 \times 10^4 \text{ mL}$

  e. $4.48 \text{ lb} \times \dfrac{453.59 \text{ g}}{1 \text{ lb}} = 2.03 \times 10^3 \text{ g}$

  f. $550 \text{ mL} \times \dfrac{1 \text{ L}}{1000 \text{ mL}} \times \dfrac{1.0567 \text{ qt}}{1 \text{ L}} = 0.58 \text{ qt}$

150. Assuming exactly 6 gross, 864 pencils

152. a.   Celsius temperature = $(175 - 32)/1.80 = 79.4$ °C
    kelvin temperature = $79.4 + 273 = 352$ K

  b. $255 - 273 = -18$ °C

  c. $(-45 - 32)/1.80 = -43$ °C

  d. $1.80(125) + 32 = 257$ °F

154. $85.5 \text{ mL} \times \dfrac{0.915 \text{ g}}{1 \text{ mL}} = 78.2 \text{ g}$

156. $m = 155 \text{ lb} \times \dfrac{453.59 \text{ g}}{1 \text{ lb}} = 7.031 \times 10^4 \text{ g}$

    $v = 4.2 \text{ ft}^3 \times \left(\dfrac{12 \text{ in}}{1 \text{ ft}}\right)^3 \times \left(\dfrac{2.54 \text{ cm}}{1 \text{ in}}\right)^3 = 1.189 \times 10^5 \text{ cm}^3$

$$d = \frac{7.031 \times 10^4 \text{ g}}{1.189 \times 10^5 \text{ cm}^3} = 0.59 \text{ g/cm}^3$$

158. a. 23 °F
     b. 32 °F
     c. -321 °F
     d. -459 °F
     e. 187 °F
     f. -459 °F

Chapter 3   Matter and Energy

2. solid, liquid, gas (vapor)

4. Liquids

6. gaseous

8. stronger

10. Gases are easily compressed into smaller volumes, whereas solids and liquids are not. Since a gaseous sample consists mostly of empty space, it is this empty space which is compressed when pressure is applied to a gas.

12. chemical

14. the substance reacts with iron(II) sulfate

16. Electrolysis is the passage of an electrical current through a substance or solution to force a chemical reaction to occur. Electrolysis causes chemical changes to take place that would ordinarily not take place on their own. When an electrical current is passed through water, the current causes the water molecules to break down into their constituent elements (hydrogen gas and oxygen gas).

18.
    a. chemical;   the scorch represents the oxidation of the material

    b. physical;   the gas in the tires decreases in volume with temperature

    c. chemical;   tarnish on silver is caused by reaction of the silver with sulfur or oxygen

    d. chemical;   the ethyl alcohol in the wine is oxidized to acetic acid

    e. chemical;   the oven cleaner contains sodium hydroxide which converts greases in the oven into soaps

    f. chemical;   an ordinary flashlight battery is constructed with a zinc casing which serves as one of the electrodes. As the battery discharges, the zinc is oxidized.

    g. chemical;   the acids attack the calcium phosphate matrix of the teeth

    h. chemical;   the charring represents the breakdown of the sugar

    i. chemical;   iron in the blood catalyzes the decomposition of the hydrogen peroxide into oxygen gas (and water)

    j. physical;   this is just a change in state; carbon dioxide, only in the gaseous state, is still present after the sublimation

Chapter 3   Matter and Energy   17

k. chemical; chlorine is an oxidizing agent and can change the chemical nature of the dyes in the fabrics

20. element

22. Compounds

24. Typically, the properties of a compound and the elements that constitute it are very different. Consider the properties of liquid *water* and the hydrogen and oxygen gases from which the water was prepared. Consider the properties of *sodium chloride* (table salt) and the sodium metal and chlorine gas from which it might have been prepared.

26. a variable

28. solutions: window cleaner, shampoo, rubbing alcohol
    mixtures: salad dressing, jelly beans, the change in my pocket

30. a. mixture

    b. mixture

    c. mixture

    d. pure substance

32. a. homogeneous

    b. heterogeneous

    c. heterogeneous

    d. homogeneous

    e. the paper itself is basically homogeneous in appearance

34. Consider a mixture of salt (sodium chloride) and sand. Salt is soluble in water, sand is not. The mixture is added to water and stirred to dissolve the salt, and is then filtered. The salt solution passes through the filter, the sand remains on the filter. The water can then be evaporated from the salt.

36. The solution is heated to vaporize (boil) the water. The water vapor is then cooled so that it condenses back to the liquid state, and the liquid is collected. After all the water is vaporized from the original sample, pure sodium chloride will remain. The process consists of physical changes.

38. the calorie

40. As the steam is cooled from 150 °C to 100 °C, the molecules of vapor gradually slow down as they lose kinetic energy. At 100 °C, the steam

18 Chapter 3   Matter and Energy

condenses into liquid water, and the temperature remains at 100 °C until all the steam has condensed. As the liquid water cools, the molecules in the liquid move more and more slowly as they lose kinetic energy. At 0 °C, the liquid water freezes.

42. temperature

44. $526 \text{ J} \times \dfrac{55 \text{ °C}}{17 \text{ °C}} = 1.7 \times 10^3 \text{ J}$

46. Since 1.000 cal = 4.184 J, then 1.000 kcal = 4.184 kJ

   a. $462.4 \text{ kJ} \times \dfrac{1 \text{ kcal}}{4.184 \text{ kJ}} = 110.5 \text{ kcal}$

   b. $18.28 \text{ kJ} \times \dfrac{1 \text{ kcal}}{4.184 \text{ kJ}} = 4.369 \text{ kcal}$

   c. $1.014 \text{ kJ} \times \dfrac{1 \text{ kcal}}{4.184 \text{ kJ}} = 0.2424 \text{ kcal}$

   d. $190.5 \text{ kJ} \times \dfrac{1 \text{ kcal}}{4.184 \text{ kJ}} = 45.53 \text{ kcal}$

48. a. $12.30 \text{ kcal} \times \dfrac{1000 \text{ cal}}{1 \text{ kcal}} = 12{,}300 \text{ cal} \ (1.230 \times 10^4 \text{ cal})$

   b. $290.4 \text{ kcal} \times \dfrac{1000 \text{ cal}}{1 \text{ kcal}} = 290{,}400 \text{ cal} \ (2.904 \times 10^5 \text{ cal})$

   c. $940{,}000 \text{ kcal} \times \dfrac{1000 \text{ cal}}{1 \text{ kcal}} = 940{,}000{,}000 \text{ cal} \ (9.4 \times 10^8 \text{ cal})$

   d. $4201 \text{ kcal} \times \dfrac{1000 \text{ cal}}{1 \text{ kcal}} = 4{,}201{,}000 \text{ cal} \ (4.201 \times 10^6 \text{ cal})$

50. a. $45.62 \text{ kcal} \times \dfrac{4.184 \text{ kJ}}{1 \text{ kcal}} = 190.9 \text{ kJ}$

   b. $72.94 \text{ kJ} \times \dfrac{1 \text{ kcal}}{4.184 \text{ kJ}} = 17.43 \text{ kcal}$

   c. $2.751 \text{ kJ} \times \dfrac{1 \text{ kcal}}{4.184 \text{ kJ}} \times \dfrac{1000 \text{ cal}}{1 \text{ kcal}} = 657.5 \text{ cal}$

   d. $5.721 \text{ kcal} \times \dfrac{4.184 \text{ kJ}}{1 \text{ kcal}} \times \dfrac{1000 \text{ J}}{1 \text{ kJ}} = 2.394 \times 10^4 \text{ J}$

52. Heat = mass × specific heat capacity × temperature change

   specific heat capacity = Heat/(mass × temperature change)

   72.4 kJ = 72,400 J

$$\text{specific heat capacity} = \frac{72{,}400 \text{ J}}{(952 \text{ g})(10.7 \text{ °C})} = 7.11 \text{ J/g °C}$$

54. Table 3.2 gives the specific heat of silver as 0.24 J/g °C

    Temperature increase = 15.2 −12.0 = 3.2 °C

    1.25 kJ = 1250 J

    1250 J = (mass of silver) × 0.24 J/g °C × 3.2 °C

    mass of silver = 1627 g = 1.6×10^3 g silver

56. Table 3.2 gives the specific heat capacity of iron as 0.45 J/g °C.

    Heat = mass × specific heat capacity × temperature change

    Temperature change = 75.5 − 40.1 = 35.4 °C

    Heat = 852.2 g × 0.45 J/g °C × 35.4 °C = 1.4×10^4 J

58. $0.13 \dfrac{\text{J}}{\text{g °C}} \times \dfrac{1 \text{ cal}}{4.184 \text{ J}} = 0.031 \dfrac{\text{cal}}{\text{g °C}}$

60. Specific heat capacities are given in Table 3.2

    for gold, 25.0 g × 0.13 J/g °C × 20. °C = 65 J

    for mercury, 25.0 g × 0.14 J/g °C × 20. °C = 70. J

    for carbon, 25.0 g × 0.71 J/g °C × 20. °C = 360 J

62. 1251 J = 35.2 g × (specific heat capacity) × 25.0 °C

    specific heat capacity = 1.42 J/g °C

64. Since *X* is a pure substance, the fact that two different solids form when electrical current is passed indicates that *X* must be a compound.

66. Since vaporized water is still the *same substance* as solid water, no chemical reaction has occurred. Sublimation is a physical change.

68. 2.5 kg of water = 2,500 g

    Temperature change = 55.0 − 18.5 = 36.5 °C

    2500 g × 4.184 J/g °C × 36.5 °C = 3.8×10^5 J

70. No calculation is necessary: aluminum will lose more heat because it has the higher specific heat capacity.

Chapter 3   Matter and Energy

72. For any substance, $Q = m \times s \times \Delta T$. The quantity of heat gained by the water in this experiment must equal the total heat lost by the metals (i.e., the *sum* of the amount of heat lost by the iron and the amount of heat lost by the aluminum).

$(m \times s \times \Delta T)_{water} = (m \times s \times \Delta T)_{iron} + (m \times s \times \Delta T)_{aluminum}$

If T_f represents the final temperature reached by this system, then

$[m \times s \times (T_f - 22.5\ °C)]_{water}$

$= [m \times s \times (100\ °C - T_f)]_{iron} + [m \times s \times (100\ °C - T_f)]_{aluminum}$

$[97.3\ g \times 4.184\ J/g\ °C \times (T_f - 22.5\ °C)]$

$= [10.00\ g \times 0.45\ J/g\ °C \times (100\ °C - T_f)]$

$+ [5.00\ g \times 0.89\ J/g\ °C \times (100\ °C - T_f)]$

$407.1(T_f - 22.5) = 4.50(100 - T_f) + 4.45(100 - T_f) = 8.95(100 - T_f)$

$407.1 T_f - 9160. = 895 - 8.95 T_f$

$416.1 T_f = 10,055$ which gives $T_f = 24.2\ °C$

74. $(m \times s \times \Delta T)_{water} = (m \times s \times \Delta T)_{iron}$

Let T_f represent the final temperature reached by the system

$[75\ g \times 4.184\ J/g\ °C \times (T_f - 20.)] = [25.0\ g \times 0.45\ J/g\ °C \times (85 - T_f)]$

$314(T_f - 20.) = 11.4(85 - T_f)$

$314 T_f - 6280 = 961 - 11.3 T_f$

$325 T_f = 7241$

$T_f = 22.3\ °C = 22\ °C$

76. far apart

78. chemical

80. chemical

82. electrolysis

84. a. heterogeneous

    b. heterogeneous

    c. heterogeneous (unless you work hard to get all the lumps out!)

d.   although strictly heterogenous, it may appear homogeneous

e.   heterogeneous

86. 9.0 J (It requires twice as much heat to warm twice as large a sample over the same temperature interval.)

88. a. $44.21 \text{ cal} \times \dfrac{4.184 \text{ J}}{1 \text{ cal}} = 185.0 \text{ J}$

b. $162.4 \text{ cal} \times \dfrac{4.184 \text{ J}}{1 \text{ cal}} = 679.5 \text{ J}$

c. $3.721 \times 10^3 \text{ cal} \times \dfrac{4.184 \text{ J}}{1 \text{ cal}} = 1.557 \times 10^4 \text{ J}$

d. $146.2 \text{ kcal} \times \dfrac{1000 \text{ cal}}{1 \text{ kcal}} \times \dfrac{4.184 \text{ J}}{1 \text{ cal}} = 6.117 \times 10^5 \text{ J}$

90. a. $5.442 \times 10^4 \text{ J} \times \dfrac{1 \text{ kJ}}{1000 \text{ J}} = 54.42 \text{ kJ}$

b. $5.442 \times 10^4 \text{ J} \times \dfrac{1 \text{ cal}}{4.184 \text{ J}} = 1.301 \times 10^4 \text{ cal}$

c. $352.6 \text{ kcal} \times \dfrac{4.184 \text{ kJ}}{1 \text{ kcal}} = 1475 \text{ kJ}$

d. $17.24 \text{ kJ} \times \dfrac{1 \text{ kcal}}{4.184 \text{ kJ}} = 4.120 \text{ kcal}$

92. The specific heat capacity of water is 4.184 J/g °C.
Temperature increase = 39 − 25 = 14 °C.
75 g × 4.184 J/g °C × 14 °C = 4400 J (to 2 significant figures)

94. For any substance, $Q = m \times s \times \Delta T$. The basic calculation for each of the substances is the same (specific heat capacities are found in Table 3.2)

Heat required = 150. g × (specific heat capacity) × 11.2 °C

Substance	Specific Heat Capacity	Heat Required
water (*l*)	4.184 J/g °C	7.03×10^3 J
water (*s*)	2.03 J/g °C	3.41×10^3 J
water (*g*)	2.0 J/g °C	3.4×10^3 J
aluminum	0.89 J/g °C	1.5×10^3 J
iron	0.45 J/g °C	7.6×10^2 J
mercury	0.14 J/g °C	2.4×10^2 J
carbon	0.71 J/g °C	1.2×10^3 J
silver	0.24 J/g °C	4.0×10^2 J
gold	0.13 J/g °C	2.2×10^2 J

96. Heat = mass × specific heat capacity × temperature change

The temperature change is 112.1 − 25.0 = 87.1 °C

1.351 kJ = 1351 J

$$s = \frac{1351 \text{ J}}{125 \text{ g} \times 87.1 \text{ °C}} = 0.124 \text{ J/g °C}$$

Cumulative Review: Chapters 1, 2, and 3

2. By now, after having covered three chapters in this book, it is hoped that you have adopted an "active" approach to your study of chemistry. You may have discovered (perhaps through a disappointing grade on a quiz (though we hope not), that you really have to get involved with chemistry. You can't just sit and take notes, or just look over the solved examples in the textbook. You have to learn to solve problems. You have to learn how to interpret problems, and how to reduce them to the simple mathematical relationships you have studied. Whereas in some courses you might get by on just giving back on exams the facts or ideas presented in class, in chemistry you have to be able to extend and synthesize what has been discussed, and to apply the material to new situations. Don't get discouraged if this is difficult at first: it's difficult for everyone at first.

4. It is difficult sometimes for students (especially beginning students) to understand why certain subjects are required for a given college major. The faculty of your major department, however, have collectively many years of experience in the subject in which you have chosen to specialize. They really do know what courses will be helpful to you in the future. They may have had trouble with the same courses that now give you trouble, but they realize that all the work will be worth it in the end. Some courses you take, particularly in your major field itself, have obvious and immediate utility. Other courses, often times chemistry included, are provided to give you a general background knowledge, which may prove useful in understanding your own major or other subjects related to your major. In perhaps a burst of bravado, chemistry has been called "the central science" by one team of textbook authors. This moniker is very true however: in order to understand biology, physics, nutrition, farming, home economics, or whatever(it helps to have a general background in chemistry.

6. Whenever a scientific measurement is made, we always employ the instrument or measuring device we are using to the limits of its precision. On a practical basis, this usually means that we *estimate* our reading of the last significant figure of the measurement. An example of the uncertainty in the last significant figure is given for measuring the length of a pin in the text in Figure 2.5. Scientists appreciate the limits of experimental techniques and instruments, and always assume that the last digit in a number representing a measurement has been estimated. Since the last significant figure in every measurement is assumed to be estimated, it is never possible to exclude uncertainty from measurements. The best we can do is to try to improve our techniques and instruments so that we get more significant figures for our measurements.

8. Dimensional analysis is a method of problem solving which pays particular attention to the units of measurements and uses these units as if they were algebraic symbols that multiply, divide, and cancel. Consider the following example. A dozen of eggs costs $1.25. Suppose we want to know how much one egg costs, and also how much three dozens of

eggs will cost. To solve these problems, we need to make use of two equivalence statements:

$$1 \text{ dozen of eggs} = 12 \text{ eggs}$$

$$1 \text{ dozen of eggs} = \$1.25$$

The first of these equivalence statements is obvious: everyone knows that 12 eggs is "equivalent" to one dozen. The second statement also expresses an equivalence: if you give the grocer $1.25, he or she will give you a dozen of eggs. From these equivalence statements, we can construct the conversion factors we need to answer the two questions. For the first question (what does one egg cost) we can set up the calculation as follows

$$\frac{\$1.25}{12 \text{ eggs}} = \$0.104 = \$0.10$$

as the cost of one egg. Similarly, for the second question (the cost of 3 dozens of eggs), we can set up the conversion as follows

$$3 \text{ dozens} \times \frac{\$1.25}{1 \text{ dozen}} = \$3.75$$

as the cost of three dozens of eggs. See section 2.6 of the text for how we construct conversion factors from equivalence statements.

10. Defining what scientists mean by "matter" often seems circular to students. Scientists say that matter is something that "has mass and occupies space", without ever really explaining what it means to "have mass" or to "occupy space"! The concept of matter is so basic and fundamental, that it becomes difficult to give a good textbook definition other than to say that matter is the "stuff" of which everything is made. Matter can be classified and subdivided in many ways, depending on what we are trying to demonstrate.
    On the most fundamental basis, all matter is composed of tiny particles (such as protons, electrons, neutrons, and the other subatomic particles). On one higher level, these tiny particles are combined in a systematic manner into units called atoms. Atoms in turn may be combined to constitute molecules. And finally, large groups of molecules may be placed together to form a bulk sample of substance that we can see.
    Matter can also be classified as to the physical state a particular substance happens to take. Some substances are solids, some are liquids, and some are gases. Matter can also be classified as to whether it is a pure substance (one type of molecule) or a mixture (more than one type of molecule), and furthermore whether a mixture is homogeneous or heterogeneous.

12. Chemists tend to give a functional definition of what they mean by an "element": an element is a fundamental substance that cannot be broken down into any simpler substances by chemical methods. Compounds, on the other hand, can be broken down into simpler substances (the elements of

which the compound is composed). For example, sulfur and oxygen are both elements (sulfur occurs as S_8 molecules and oxygen as O_2 molecules). When sulfur and oxygen are placed together and heated, the compound sulfur dioxide (SO_2) forms. When we analyze the sulfur dioxide produced, we notice that each and every molecule consists of one sulfur atom and two oxygen atoms, and on a mass basis, consists of 50% each of sulfur and oxygen. We describe this by saying that sulfur dioxide has a constant composition. The fact that a given compound has constant composition is usually expressed in terms of the mass percentages of the elements present in the compound, but realize that the reason the mass percentages are constant is because of a constant number of atoms of each type present in the compound's molecules. If a scientist anywhere in the universe analyzed sulfur dioxide, he or she would find the same composition: if a scientist finds something that does not have the same composition, then the substance cannot be sulfur dioxide.

14. Scientists define energy as "the capacity to do work". As with trying to define "matter" earlier in this chapter, energy is such a fundamental concept that it is hard to define (what is "work"?). Although the *SI* unit of energy is the *joule*, until relatively recently, energies were more commonly given in terms of the *calorie*: one calorie is defined to be the amount of heat required to raise the temperature of one gram of water by one Celsius degree. The calorie is a "working" definition, and we can more easily appreciate this amount of energy. In terms of the *SI* unit, 1 calorie = 4.184 joule, so it takes 4.184 J to raise the temperature of one gram of water by one Celsius degree.

 The specific heat capacity of a substance, in general, is the amount of energy required to raise the temperature of one gram of a substance by one Celsius degree. Therefore, the specific heat capacity of water must be 1.000 cal/g°C or 4.184 J/g°C. To see how specific heat capacities may be used to calculate the energy change for a process, consider this example: How much energy is required to warm 25.0 g of water from 15.1°C to 35.2°C? The specific heat capacity of water is 1.000 cal/g°C: this is the quantity of energy required to raise the temperature of only *one* gram of water by only *one* Celsius degree. In this example, we are raising the temperature of 25.0 g of water, and we are raising the temperature by (35.2 - 15.1) = 20.1°C. So, using the specific heat capacity as a conversion factor, we can say

$$\text{energy required} = \left(\frac{1.00 \text{ cal}}{\text{g °C}}\right) \times (25.0 \text{ g}) \times (20.1\text{°C}) = 503 \text{ cal}$$

Notice how the units of g and °C cancel, leaving the answer in energy units only. This sort of calculation of energy change can be done for any substance, using the substance's own specific heat capacity (see Table 3.2).

16. a. $6.0 \text{ pt} \times \dfrac{1 \text{ qt}}{2 \text{ pt}} \times \dfrac{1 \text{ L}}{1.0567 \text{ qt}} = 2.8 \text{ L}$

b. $6.0 \text{ pt} \times \dfrac{1 \text{ qt}}{2 \text{ pt}} \times \dfrac{1 \text{ gal}}{4 \text{ qt}} = 0.75 \text{ gal}$

c. $5.91 \text{ yd} \times \dfrac{1 \text{ m}}{1.0936 \text{ yd}} = 5.40 \text{ m}$

d. $16.0 \text{ L} \times \dfrac{1 \text{ qt}}{0.94633 \text{ L}} \times \dfrac{32 \text{ fl. oz.}}{1 \text{ qt}} = 541 \text{ fl. oz.}$

e. $5.25 \text{ L} \times \dfrac{1 \text{ gal}}{3.7854 \text{ L}} = 1.39 \text{ gal}$

f. $62.5 \text{ mi} \times \dfrac{1 \text{ km}}{0.62137 \text{ mi}} = 101 \text{ km}$

g. $8.25 \text{ m} \times \dfrac{1.0936 \text{ yd}}{1 \text{ m}} \times \dfrac{36 \text{ in}}{1 \text{ yd}} = 325 \text{ in}$

h. $4.25 \text{ kg} \times \dfrac{2.2046 \text{ lb}}{1 \text{ kg}} = 9.37 \text{ lb}$

i. $88.5 \text{ cm} \times \dfrac{10 \text{ mm}}{1 \text{ cm}} = 885 \text{ mm}$

j. $4.21 \text{ in} \times \dfrac{2.54 \text{ cm}}{1 \text{ in}} = 10.7 \text{ cm}$

18. $t_F = 1.80(t_C) + 32 \qquad t_C = (t_F - 32)/1.80 \qquad t_K = t_C + 273$

    a. $1.80(-50.1°C) + 32 = -58.2°F$

    b. $(-30.7°C - 32)/1.80 = -34.8°C$

    c. $541 \text{ K} - 273 = 268°C$

    d. $221°C + 273 = 494 \text{ K}$

    e. $351 \text{ K} - 273 = 78°C$

    $1.80(78°C) + 32 = 172.4 = 172°F$

f. (72°F − 32)/1.80 = 22.2°C

22.2°C + 273 = 295.2 = 295 K

20. a. $459 \text{ J} \times \dfrac{1 \text{ cal}}{4.184 \text{ J}} = 109.7 = 110. \text{ cal}$

b. $7{,}031 \text{ cal} \times \dfrac{4.184 \text{ J}}{1 \text{ cal}} \times \dfrac{1 \text{ kJ}}{1000 \text{ J}} = 29.42 \text{ kJ}$

c. $55.31 \text{ kJ} = 55{,}310 \text{ J} = 5.531 \times 10^4 \text{ J}$

d. $78.3 \text{ kcal} \times \dfrac{4.184 \text{ kJ}}{1 \text{ kcal}} = 327.6 = 328 \text{ kJ}$

e. $4{,}541 \text{ cal} = 4.541 \text{ kcal}$

f. $84.1 \text{ kJ} \times \dfrac{1 \text{ kcal}}{4.184 \text{ kJ}} = 20.1 \text{ kcal}$

Chapter 4   Chemical Foundations: Elements, Atoms, and Ions

2. The alchemists discovered several previously unknown elements (mercury, sulfur, antimony) and were the first to prepare several common acids.

4. There are 112 elements presently known; of these 88 occur naturally and 24 are manmade. Table 4.1 lists the most common elements on the earth.

6. The four most abundant elements in living creatures are, respectively, oxygen, carbon, hydrogen, and nitrogen (see Table 4.2). In the nonliving world, the most abundant elements are, respectively, oxygen, silicon, aluminum, and iron (see Table 4.1).

8. Sb (antimony)

    Cu (copper)

    Au (gold)

    Pb (lead)

    Hg (mercury)

    K (potassium)

    Ag (silver)

    Na (sodium)

    Sn (tin)

    W (tungsten)

    Fe (iron)

10. a. Al
    b. Fe
    c. F
    d. Ca
    e. Au
    f. Hg

12. Ir   iridium

    Ta   tantalum

    Bi   bismuth

    Pu   plutonium

    Fr   francium

    At   astatine

Chapter 4   Chemical Foundations: Elements, Atoms, and Ions

14. a. copper
    b. cobalt
    c. calcium
    d. carbon
    e. chromium
    f. cesium
    g. chlorine
    h. cadmium

16. According to Dalton, a given compound is always made up of the same number and type of atoms, and so the composition of the compound on a mass percentage basis will always be the same, no matter what the source of the compound is. For example, water molecules from the Atlantic Ocean and the Pacific Ocean all contain two hydrogen atoms bonded to one oxygen atom.

18. According to Dalton, all atoms of the same element are *identical*; in particular, every atom of a given element has the same *mass* as every other atom of that element. If a given compound always contains the *same relative numbers* of atoms of each kind, and those atoms always have the *same masses*, then it follows that the compound made from those elements would always contain the same relative masses of its elements.

20. a. C_6H_6
    b. N_2O_4
    c. $CaCl_2$
    d. $FeBr_3$
    e. $NaNO_3$
    f. Ca_3N_2

22. a. False; Rutherford's bombardment experiments with metal foil suggested that the alpha particles were being deflected by coming near a *dense, positively charged* atomic nucleus.

    b. False; The proton and the electron have opposite charges, but the mass of the electron is *much smaller* than the mass of the proton.

    c. True

24. protons

26. neutron; electron

28. electrons

Chapter 4   Chemical Foundations: Elements, Atoms, and Ions

30. False; the mass number represents the total number of protons and neutrons in the nucleus.

32. mass

34. Atoms of the same element (i.e., atoms with the same number of protons in the nucleus) may have different numbers of neutrons, and so will have different masses.

36. a. 32

b. 30

c. 24

d. 74

e. 38

f. 27

g. 4

h. 3

38. a. $^{12}_{5}B$

b. $^{15}_{7}N$

c. $^{35}_{17}Cl$

d. $^{235}_{92}U$

e. $^{14}_{6}C$

f. $^{31}_{15}P$

40. a. 6 protons, 6 neutrons, 6 electrons

b. 27 protons, 33 neutrons, 27 electrons

c. 17 protons, 20 neutrons, 17 electrons

d. 55 protons, 77 neutrons, 55 electrons

e. 92 protons, 146 neutrons, 92 electrons

f. 26 protons, 30 neutrons, 26 electrons

42.

element	neutrons	atomic number	mass number	symbol
nitrogen	6	7	13	$^{13}_{7}N$
nitrogen	7	7	14	$^{14}_{7}N$
lead	124	82	206	$^{206}_{82}Pb$
iron	31	26	57	$^{57}_{26}Fe$
krypton	48	36	84	$^{84}_{36}Kr$

44. Elements with similar chemical properties are aligned *vertically* in families known as *groups*.

46. Metallic elements are found towards the *left* and *bottom* of the periodic table; there are far more metallic elements than there are nonmetals.

48. hydrogen, nitrogen, oxygen, fluorine, chlorine, plus all the group 8 elements (noble gases)

50. The metalloids are the elements found on either side of the "stairstep" region that is marked on most periodic tables. The metalloid elements show some properties of both metals and nonmetals.

52. a. Group 7; halogens

    b. Group 2; alkaline earth elements

    c. Group 1; alkali metals

    d. Group 1; alkali metals

    e. Group 8; noble gases

    f. Group 1; alkali metals

    g. Group 8; noble gases

Chapter 4 Chemical Foundations: Elements, Atoms, and Ions

54.

name	symbol	atomic number	group number	metal/nonmetal
rubidium	Rb	37	1	metal
germanium	Ge	32	4	metalloid
magnesium	Mg	12	2	metal
titanium	Ti	22	-	transition metal
iodine	I	53	7	nonmetal

56. Most of the elements are too reactive to be found in the uncombined form in nature, and are found only in compounds.

58. These elements are found *uncombined* in nature and do not readily react with other elements. For many years it was thought that these elements formed no compounds at all, although this has now been shown to be untrue.

60. Diatomic gases: H_2, N_2, O_2, Cl_2, and F_2
    Monatomic gases: He, Ne, Kr, Xe, Rn, and Ar

62. chlorine

64. diamond

66. electrons

68. 2+

70. *-ide*

72. nonmetallic

74.
  a. Co^{2+}:   27 protons   25 electrons   CoO
  b. Co^{3+}:   27 protons   24 electrons Co_2O_3
  c. Cl^-:   17 protons   18 electrons $CaCl_2$
  d. K^+:   19 protons   18 electrons K_2O
  e. S^{2-}:   16 protons   18 electrons   CaS
  f. Sr^{2+}:   38 protons   36 electrons   SrO
  g. Al^{3+}:   13 protons   10 electrons Al_2O_3
  h. P^{3-}:   15 protons   18 electrons Ca_3P_2

Chapter 4   Chemical Foundations: Elements, Atoms, and Ions   33

76. a. $2e^-$

    b. $2e^-$

    c. $2e^-$

    d. $3e^-$

    e. $1e^-$

    f. $3e^-$

78. a. Ra^{2+} (element 88, Ra, is in Group 2)

    b. Te^{2-} (element 52, Te, is in Group 6)

    c. I^- (element 53, I, is in Group 7)

    d. Fr^+ (element 87, Fr, is in Group 1)

    e. At^- (element 85, At, is in Group 7)

    f. no ion is likely   (element 86, Rn, is a noble gas)

80. Sodium chloride is an *ionic* compound, consisting of Na^+ and Cl^- *ions*. When NaCl is dissolved in water, these ions are *set free*, and can move independently to conduct the electrical current. Sugar crystals, although they may visually *appear* similar contain *no* ions. When sugar is dissolved in water, it dissolves as uncharged *molecules*. There are no electrically charged species present in a sugar solution to carry the electrical current.

82. The total number of positive charges must equal the total number of negative charges so that there will be *no net charge* on the crystals of an ionic compound. A macroscopic sample of compound must ordinarily not have any net charge.

84. a. One 2+ ion is exactly balanced by one 2- ion: BaO

    b. Three 1+ ions are needed to balance one 3- ion: K_3P

    c. Two 2+ ions are required to balance one 4- ion: Ca_2C

    d. The smallest common multiple of 3 and 2 is 6; two 3+ ions are balanced by three 2- ions: Al_2S_3

    e. The smallest common multiple of 2 and 3 is 6; three 2+ ions are balanced by two 3- ions: Sr_3P_2

    f. One 1+ ion is balanced by one 1- ion: NaI

    g. One 3+ ion is balanced by three 1- ions: $CoCl_3$

    h. One 4+ ion is balanced by four 1- ions: $SnBr_4$

86. a. 7; halogens
     b. 8; noble gases
     c. 2; alkaline earth elements
     d. 2; alkaline earth elements
     e. 4
     f. 6; (the members of group 6 are sometimes called the chalcogens)
     g. 8; noble gases
     h. 1; alkali metals

88.
	element	symbol	atomic number
Group 3	boron	B	5
	aluminum	Al	13
	gallium	Ga	31
	indium	In	49
Group 5	nitrogen	N	7
	phosphorus	P	15
	arsenic	As	33
	antimony	Sb	51
Group 8	helium	He	2
	neon	Ne	10
	argon	Ar	18
	krypton	Kr	36

90. Most of the mass of an atom is concentrated in the nucleus: the *protons* and *neutrons* which constitute the nucleus have similar masses, and these particles are nearly two thousand times heavier than electrons. The chemical properties of an atom depend on the number and location of the *electrons* it possesses. Electrons are found in the outer regions of the atom, and are the particles most likely to be involved in interactions between atoms.

92. $C_6H_{12}O_6$

94. a. 29 protons; 34 neutrons; 29 electrons
     b. 35 protons; 45 neutrons; 35 electrons
     c. 12 protons; 12 neutrons; 12 electrons

96. The chief use of gold in ancient times was as *ornamentation*, whether in statuary or in jewelry. Gold possesses an especially beautiful luster, and since it is relatively soft and malleable, it could be worked finely by

artisans; among the metals, gold is particularly inert to attack by most substances in the environment.

98.
 a. I
 b. Si
 c. W
 d. Fe
 e. Cu
 f. Co

100.
 a. Br
 b. Bi
 c. Hg
 d. V
 e. F
 f. Ca

102.
 a. osmium
 b. zirconium
 c. rubidium
 d. radon
 e. uranium
 f. manganese
 g. nickel
 h. bromine

104.
 a. CO_2
 b. $AlCl_3$
 c. $HClO_4$
 d. SCl_6

106.
 a. $^{13}_{6}C$
 b. $^{13}_{6}C$
 c. $^{13}_{6}C$

d. $^{44}_{19}K$

e. $^{41}_{20}Ca$

f. $^{35}_{19}K$

108.

	symbol	number of protons	number of neutrons	mass number
a.	$^{41}_{20}Ca$	20	21	41
b.	$^{55}_{25}Mn$	25	30	55
c.	$^{109}_{47}Ag$	47	62	109
d.	$^{45}_{21}Sc$	21	24	45

Chapter 5   Nomenclature

2. compounds that contain a metal and a nonmetal; compounds containing two nonmetals

4. cation

6. The substance "sodium chloride" consists of an extended lattice array of sodium ions, Na^+, and chloride ions, Cl^-. Each sodium ion is surrounded by several chloride ions, and each chloride ion is surrounded by several sodium ions. We write the formula as NaCl to indicate the relative number of each ion in the substance.

8. Roman numeral

10.
   a. potassium chloride
   b. barium oxide
   c. rubidium sulfide
   d. sodium phosphide
   e. aluminum fluoride
   f. magnesium nitride
   g. calcium iodide
   h. radium chloride

12.
   a. incorrect; silver chloride is AgCl
   b. correct
   c. incorrect; sodium oxide would be Na_2O
   d. incorrect; barium chloride is $BaCl_2$
   e. incorrect; strontium oxide is SrO

14.
   a. Since each iodide ion has a 1- charge, the iron ion must have a 3+ charge: the name is iron(III) iodide.
   b. Since each chloride ion has a 1- charge, the manganese must have a 2+ charge: the name is manganese(II) chloride.
   c. Since the oxide ion has a 2- charge, the mercury ion must have a 2+ charge: mercury(II) oxide.
   d. Since the oxide ion has a 2- charge, the copper atoms must each have a 1+ charge: the name is copper(I) oxide.
   e. Since the oxide ion has a 2- charge, the copper ion must have a 2+ charge: copper(II) oxide.
   f. Since each bromide ion has a 1- charge, the tin ion must have a 4+ charge: tin(IV) bromide.

38 Chapter 5   Nomenclature

16. a. Since oxide ions have a 2- charge the lead ion must have a 4+ charge: the name is plumb*ic* oxide.

   b. Since bromide ions have a 1- charge, the tin ion must have a 2+ charge: the name is stann*ous* bromide.

   c. Since the sulfide ion has a 2- charge, the two copper ions must each have a 1+ charge: the name is cupr*ous* sulfide.

   d. Since the iodide ion has a 1- charge, the copper ion must have a 1+ charge: the name is cupr*ous* iodide.

   e. Since iodide ions have a 1- charge, each mercury must have a 1+ charge: the name is mercur*ous* iodide.

   f. Since fluoride ions have a 1- charge, the chromium ion must have a 3+ charge: the name is chrom*ic* fluoride.

18. a. germanium tetrahydride

   b. dinitrogen tetrabromide

   c. diphosphorus pentasulfide

   d. selenium dioxide

   e. ammonia (nitrogen trihydride)

   f. silicon dioxide

20. a. diboron hexahydride - nonionic (common name: *diborane*)

   b. calcium nitride - ionic

   c. carbon tetrabromide - nonionic

   d. silver sulfide - ionic

   e. copper(II) chloride, cupric chloride - ionic

   f. chlorine monofluoride - nonionic

22. a. radium chloride - ionic

   b. selenium dichloride - nonionic

   c. phosphorus trichloride - nonionic

   d. sodium phosphide - ionic

   e. manganese(II) fluoride - ionic

   f. zinc oxide - ionic

24. An oxyanion is a polyatomic anion containing oxygen combined with another element. The following are the oxyanions of bromine:

    BrO^- hypobromite
    BrO_2^- bromite
    BrO_3^- bromate
    BrO_4^- perbromate

26. perchlorate, ClO_4^-

28. hypobromite

    IO_3^-

    periodate

    OI^- or IO^-

30. a. NO_3^-
    b. NO_2^-
    c. NH_4^+
    d. CN^-

32. a. $MgCl_2$
    b. $Ca(ClO)_2$
    c. $KClO_3$
    d. $Ba(ClO_4)_2$

34. a. ammonium
    b. dihydrogen phosphate
    c. sulfate
    d. hydrogen sulfite (also called *bi*sulfite)
    e. perchlorate
    f. iodate

36. a. ammonium sulfate
    b. potassium perchlorate
    c. iron(III) sulfate, ferric sulfate
    d. calcium phosphate
    e. calcium hydroxide
    f. potassium carbonate

Chapter 5   Nomenclature

38. oxygen (commonly referred to as *oxyacids*)

40. a. hypochlorous acid
    b. sulfurous acid
    c. bromic acid
    d. hypoiodous acid
    e. perbromic acid
    f. hydrosulfuric acid
    g. hydroselenic acid
    h. phosphorous acid

42. a. PbO_2
    b. $SnBr_2$
    c. CuS
    d. CuI
    e. Hg_2Cl_2
    f. CrF_3

44. a. CO_2
    b. SO_2
    c. N_2Cl_4
    d. CI_4
    e. PF_5
    f. P_2O_5

46. a. $Ca_3(PO_4)_2$
    b. NH_4NO_3
    c. $Al(HSO_4)_3$
    d. $BaSO_4$
    e. $Fe(NO_3)_3$
    f. $CuOH$

48. a. HCN
    b. HNO_3
    c. H_2SO_4
    d. H_3PO_4

Chapter 5   Nomenclature   41

      e.   HClO or HOCl

      f.   HBr

      g. $HBrO_2$

      h.   HF

50. a. $Mg(HSO_4)_2$

      b. $CsClO_4$

      c.   FeO

      d. H_2Te

      e. $Sr(NO_3)_2$

      f. $Sn(C_2H_3O_2)_4$

      g. $MnSO_4$

      h. N_2O_4

      i. Na_2HPO_4

      j. Li_2O_2

      k. HNO_2

      l. $Co(NO_3)_3$

52. A moist paste of NaCl would contain Na^+ and Cl^- ions in solution, and would serve as a *conductor* of electrical impulses.

54. $H \rightarrow H^+$ (hydrogen ion: a cation) $+ e^-$
     $H + e^- \rightarrow H^-$ (hydr*ide* ion: an anion)

56. missing oxyanions: IO_3^-; ClO_2^-

     missing oxyacids: $HClO_4$; HClO; $HBrO_2$

58. a.   gold(III) bromide, auric bromide

      b.   cobalt(III) cyanide, cobaltic cyanide

      c.   magnesium hydrogen phosphate

      d.   diboron hexahydride (diborane is its common name)

      e.   ammonia

      f.   silver(I) sulfate (usually called silver sulfate)

      g.   beryllium hydroxide

60. a.   ammonium carbonate

      b.   ammonium hydrogen carbonate, ammonium bicarbonate

42   Chapter 5   Nomenclature

      c. calcium phosphate
      d. sulfurous acid
      e. manganese(IV) oxide
      f. iodic acid
      g. potassium hydride

62. a. $M(C_2H_3O_2)_4$
      b. $M(MnO_4)_4$
      c. MO_2
      d. $M(HPO_4)_2$
      e. $M(OH)_4$
      f. $M(NO_2)_4$

64. M^+ compounds: MD, M_2E, M_3F

     M^{2+} compounds: MD_2, ME, M_3F_2

     M^{3+} compounds: MD_3, M_2E_3, MF

66.

$Ca(NO_3)_2$	$CaSO_4$	$Ca(HSO_4)_2$	$Ca(H_2PO_4)_2$	CaO	$CaCl_2$
$Sr(NO_3)_2$	$SrSO_4$	$Sr(HSO_4)_2$	$Sr(H_2PO_4)_2$	SrO	$SrCl_2$
NH_4NO_3	$(NH_4)_2SO_4$	NH_4HSO_4	$NH_4H_2PO_4$	$(NH_4)_2O$	NH_4Cl
$Al(NO_3)_3$	$Al_2(SO_4)_3$	$Al(HSO_4)_3$	$Al(H_2PO_4)_3$	Al_2O_3	$AlCl_3$
$Fe(NO_3)_3$	$Fe_2(SO_4)_3$	$Fe(HSO_4)_3$	$Fe(H_2PO_4)_3$	Fe_2O_3	$FeCl_3$
$Ni(NO_3)_2$	$NiSO_4$	$Ni(HSO_4)_2$	$Ni(H_2PO_4)_2$	NiO	$NiCl_2$
$AgNO_3$	Ag_2SO_4	$AgHSO_4$	AgH_2PO_4	Ag_2O	$AgCl$
$Au(NO_3)_3$	$Au_2(SO_4)_3$	$Au(HSO_4)_3$	$Au(H_2PO_4)_3$	Au_2O_3	$AuCl_3$
KNO_3	K_2SO_4	$KHSO_4$	KH_2PO_4	K_2O	KCl
$Hg(NO_3)_2$	$HgSO_4$	$Hg(HSO_4)_2$	$Hg(H_2PO_4)_2$	HgO	$HgCl_2$
$Ba(NO_3)_2$	$BaSO_4$	$Ba(HSO_4)_2$	$Ba(H_2PO_4)_2$	BaO	$BaCl_2$

68. helium

70. iodine (solid), bromine (liquid), fluorine and chlorine (gases)

72. 1−

74. 1−

76. a. $Al(13e^-) \rightarrow Al^{3+}(10e^-) + 3e^-$

      b. $S(16e^-) + 2e^- \rightarrow S^{2-}(18e^-)$

      c. $Cu(29e^-) \rightarrow Cu^+(28e^-) + e^-$

d. $F(9e^-) + e^- \rightarrow F^-(10e^-)$

e. $Zn(30e^-) \rightarrow Zn^{2+}(28e^-) + 2e^-$

f. $P(15e^-) + 3e^- \rightarrow P^{3-}(18e^-)$

78. a. Two 1+ ions are needed to balance a 2- ion, so the formula must have two Na^+ ions for each S^{2-} ion: Na_2S.

    b. One 1+ ion exactly balances a 1- ion, so the formula should have an equal number of K^+ and Cl^- ions: KCl.

    c. One 2+ ion exactly balances a 2- ion, so the formula must have an equal number of Ba^{2+} and O^{2-} ions: BaO.

    d. One 2+ ion exactly balances a 2- ion, so the formula must have an equal number of Mg^{2+} and Se^{2-} ions: $MgSe$.

    e. One 2+ ion requires two 1- ions to balance charge, so the formula must have twice as many Br^- ions as Cu^{2+} ions: $CuBr_2$.

    f. One 3+ ion requires three 1- ions to balance charge, so the formula must have three times as many I^- ions as Al^{3+} ions: AlI_3.

    g. Two 3+ ions give a total of 6+, whereas three 2- ions will give a total of 6-. The formula then should contain two Al^{3+} ions and three O^{2-} ions: Al_2O_3.

    h. Three 2+ ions are required to balance two 3- ions, so the formula must contain three Ca^{2+} ions for every two N^{3-} ions: Ca_3N_2.

80. a. incorrect. Si is the element silicon, not silver.

    b. incorrect. Co is the symbol for cobalt, not copper.

    c. incorrect. Hydrogen exists as the hydride ion in this compound.

    d. correct

    e. incorrect. P is just "phosphorus" not "phosphoric".

82. a. Since bromide ions always have a 1- charge, the cobalt ion must have a 3+ charge: the name is cobalt*ic* bromide.

    b. Since iodide ions always have a 1- charge, the lead ion must have a 4+ charge: the name is plumb*ic* iodide.

    c. Since oxide ions always have a 2- charge, and since there are three oxide ions, each iron ion must have a 3+ charge: the name is ferr*ic* oxide.

44   Chapter 5   Nomenclature

   d.   Since sulfide ions always have a 2- charge, the iron ion must have a 2+ charge: the name is ferrous sulfide.

   e.   Since chloride ions always have a 1- charge, the tin ion must have a 4+ charge: the name is stannic chloride.

   f.   Since oxide ions always have a 2- charge, the tin ion must have a 2+ charge: the name is stannous oxide.

84. a.   iron(III) acetate, ferric acetate
     b.   bromine monofluoride
     c.   potassium peroxide
     d.   silicon tetrabromide
     e.   copper(II) permanganate, cupric permanganate
     f.   calcium chromate

86. a. CO_3^{2-}
     b. HCO_3^-
     c. $C_2H_3O_2^-$
     d. CN^-

88. a.   carbonate
     b.   chlorate
     c.   sulfate
     d.   phosphate
     e.   perchlorate
     f.   permanganate

90. a. $CaCl_2$
     b. Ag_2O
     c. Al_2S_3
     d. $BeBr_2$
     e. H_2S
     f. KH
     g. MgI_2
     h. CsF

92. a. NaH_2PO_4
     b. $LiClO_4$
     c. $Cu(HCO_3)_2$
     d. $KC_2H_3O_2$
     e. BaO_2
     f. Cs_2SO_3

Cumulative Review: Chapters 4 and 5

2. How many elements could you name? While you certainly don't have to memorize all the elements, you should at least be able to give the symbol or name for the most common elements (listed in Table 4.3).

4. Dalton's atomic theory as presented in this text consists of five main postulates. Realize that although Dalton's theory was exceptional scientific thinking for its time, some of the postulates have been modified as our scientific instruments and calculational methods have become increasingly more sophisticated. The main postulates of Dalton's theory are as follows: (1) Elements are made up of tiny particles called atoms; (2) all atoms of a given element are identical; (3) although all atoms of a given element are identical, these atoms are different from the atoms of all other elements; (4) atoms of one element can combine with atoms of another element to form a compound, and such a compound will always have the same relative numbers and types of atoms for its composition; (5) atoms are merely rearranged into new groupings during an ordinary chemical reaction, and no atom is ever destroyed and no new atom is ever created during such a reaction.

6. The expression *nuclear* atom indicates that we view the atom as having a dense center of positive charge (called the nucleus) around which the electrons move through primarily empty space. Rutherford's experiment involved shooting a beam of particles at a thin sheet of metal foil. According to the then current "plum pudding" model of the atom, most of these positively charged particles should have passed right through the foil. However, Rutherford detected that a significant number of particles effectively bounced off something and were deflected backwards to the source of particles, and that other particles were deflected from the foil at large angles. Rutherford realized that his observations could be explained if the atoms of the metal foil had a small, dense, positively charged nucleus, with a significant amount of empty space between nuclei. The empty space between nuclei would allow most of the particles to pass through the atom. However, if an particle hit a nucleus head-on, it would be deflected backwards at the source. If a positively-charged particle passed near a positively charged nucleus (but did not hit the nucleus head-on), then the particle would be deflected by the repulsive forces between the positive charges. Rutherford's experiment conclusively disproved the "plum pudding" model for the atom, which envisioned the atom as a uniform sphere of positive charge, with enough negatively charged electrons scattered through the atom to balance out the positive charge.

8. Isotopes represent atoms of the same element which have different atomic masses. Isotopes are a result of the fact that atoms of a given element may have different numbers of neutrons in their nuclei. Isotopes have the same atomic number (number of protons in the nucleus) but have different mass numbers (total number of protons and neutrons in the nucleus). The different isotopes of an atom are indicated by symbolism of the form $^{A}_{Z}X$ in which Z represents the atomic number, and A the mass number, of element X. For example, $^{13}_{6}C$ represents a nuclide of carbon

with atomic number 6 (6 protons in the nucleus) and mass number 13 (reflecting 6 protons plus 7 neutrons in the nucleus). The various isotopes of an element have identical chemical properties since the chemical properties of an atom are a function of the electrons in the atom (*not* the nucleus). The physical properties of the isotopes of an element (and compounds containing those isotopes) may differ because of the difference in mass of the isotopes.

10. Most elements are too reactive to be found in nature in other than the combined form. Aside from the noble metals gold, silver, and platinum, the only other elements commonly found in nature in the uncombined state are some of the gaseous elements (such as O_2, N_2, He, Ar, etc.), and the solid nonmetals carbon and sulfur.

12. Ionic compounds typically are hard, crystalline solids with high melting and boiling points. Ionic substances like sodium chloride, when dissolved in water or when melted, conduct electrical currents: chemists have taken this evidence to mean that ionic substances consist of positively and negatively charged particles (ions). Although an ionic substance is made up of positively and negatively charged particles, there is no net electrical charge on a sample of such a substance because the total number of positive charges is balanced by an equal number of negative charges. An ionic compound could not possibly exist of just cations or just anions: there must be a balance of charge or the compound would be very unstable (like charges repel each other).

14. When naming ionic compounds, we name the positive ion (cation) first. For simple binary Type I ionic compounds, the ending *-ide* is added to the root name of the element which is the negative ion (anion). For example, for the Type I ionic compound formed between potassium and sulfur, K_2S, the name would be potassium sulfide: potassium is the cation, sulfur is the anion (with the suffix *-ide* added). Type II compounds are named by either of two systems, the "*ous-ic*" system (which is falling out of use), and the "Roman numeral" system which is preferred by most chemists. Type II compounds involve elements which form more than one stable ion, and so it is necessary to specify *which* ion is present in a given compound. For example, iron forms two types of stable ion: Fe^{2+} and Fe^{3+}. Iron can react with oxygen to form either of two stable oxides, FeO or Fe_2O_3, depending on which cation is involved. Under the Roman numeral naming system, FeO would be named iron(II) oxide to show that it contains Fe^{2+} ions; Fe_2O_3 would be named iron(III) oxide to indicate that it contains Fe^{3+} ions. The Roman numeral used in a name corresponds to the charge of the specific ion present in the compound. Under the less-favored "ous-ic" system, for an element that forms two stable ions, the ending *-ous* is used to indicated the lower-charged ion, whereas the ending *-ic* is used to indicate the higher-charged ion. FeO and Fe_2O_3 would thus be named ferr*ous* oxide and ferr*ic* oxide, respectively. The "ous-ic" system has fallen out of favor since it does not indicate the actual charge on the ion, but only that it is the lower or higher charged of the two. This can lead to confusion: for example Fe^{2+} is called ferrous ion in this system, but Cu^{2+} is called cupric ion (since there is also a Cu^+ stable ion).

16. A polyatomic ion is an ion containing more than one atom. Some common polyatomic ions you should be familiar with are listed in Table 5.4. Parentheses are used in writing formulas containing polyatomic ions to indicate unambiguously how many of the polyatomic ion are present in the formula, to make certain that there is no mistake as to what is meant by the formula. For example, consider the substance calcium phosphate. The correct formula for this substance is $Ca_3(PO_4)_2$, which indicates that three calcium ions are combined for every two phosphate ions (check the total number of positive and negative charges to see why this is so). If we did not write the parenthesis around the formula for the phosphate ion, that is, if we had written Ca_3PO_{42}, people reading this formula might think that there were 42 oxygen atoms present!

18. Acids, in general, are substances which produce protons (H^+ ions) when dissolved in water. For acids which do not contain oxygen, the prefix *hydro-* and the suffix *-ic* are used with the root name of the element present in the acid (for example: HCl, hydrochloric acid; H_2S, hydrosulfuric acid; HF, hydrofluoric acid). The nomenclature of acids whose anions contain oxygen is more complicated. A series of prefixes and suffixes is used with the name of the non-oxygen atom in the anion of the acid: these prefixes and suffixes indicate the relative (not actual) number of oxygen atoms present in the anion. Most of the elements that form oxyanions form two such anions: for example, sulfur forms sulfite ion (SO_3^{2-}) and sulfate ion (SO_4^{2-}), and nitrogen forms nitrite ion (NO_2^-) and nitrate ion (NO_3^-). For an element that forms two oxyanions, the acid containing the anions will have the ending *-ous* if the anion is the *-ite* anion and the ending *-ic* if the anion is the *-ate* anion. For example, HNO_2 is nitr*ous* acid and HNO_3 is nitr*ic* acid; H_2SO_3 is sulfur*ous* acid and H_2SO_4 is sulfur*ic* acid. The halogen elements (Group 7) each form four oxyanions, and consequently, four oxyacids. The prefix *hypo-* is used for the oxyacid that contains fewer oxygen atoms than the *-ite* anion, and the prefix *per-* is used for the oxyacid that contains more oxygen atoms than the *-ate* anion. For example,

acid	*name*	*anion*	*anion name*
HBrO	*hypo*brom*ous* acid	BrO^-	*hypo*brom*ite*
$HBrO_2$	brom*ous* acid	BrO_2^-	brom*ite*
$HBrO_3$	brom*ic* acid	BrO_3^-	brom*ate*
$HBrO_4$	*per*brom*ic* acid	BrO_4^-	*per*brom*ate*

20.

	Formula	Name	Atomic Number
a.	He	helium	2
b.	B	boron	5
c.	C	carbon	6
d.	F	fluorine	9
e.	S	sulfur	16
f.	Ba	barium	56

g. Be beryllium 4
h. O oxygen 8
i. P phosphorus 15
j. Si silicon 14

22.

		protons	neutrons	electrons
a.	$^{4}_{2}He$	2	2	2
b.	$^{37}_{17}Cl$	17	20	17
c.	$^{79}_{35}Br$	35	44	35
d.	$^{41}_{20}Ca$	20	21	20
e.	$^{40}_{20}Ca$	20	20	20
f.	$^{238}_{92}U$	92	146	92
g.	$^{235}_{92}U$	92	143	92
h.	$^{1}_{1}H$	1	0	1

24. a. K^+ (19 protons, 18 electrons)
b. Ca^{2+} (20 protons, 18 electrons)
c. N^{3-} (7 protons, 10 electrons)
d. Br^- (35 protons, 36 electrons)
e. Al^{3+} (13 protons, 10 electrons)
f. Ag^+ (47 protons, 46 electrons)
g. Cl^- (17 protons, 18 electrons)
h. H^+ (1 proton, 0 electrons)
i. H^- (1 proton, 2 electrons)
j. Na^+ (11 protons, 10 electrons)
k. O^{2-} (8 protons, 10 electrons)
l. I^- (53 protons, 54 electrons)

26. a. $FeCl_3$, iron(III) chloride, ferric chloride
    b. Cu_2S, copper(I) sulfide, cuprous sulfide
    c. $CoBr_2$, cobalt(II) bromide, cobaltous bromide
    d. Fe_2O_3, iron(III) oxide, ferric oxide
    e. AuI_3, gold(III) iodide, auric iodide
    f. Cr_2S_3, chromium(III) sulfide, chromic sulfide
    g. MnO_2, manganese(IV) oxide, manganese dioxide (archaic)
    h. CuO, copper(II) oxide, cupric oxide
    i. NiS, nickel(II) sulfide, nickelous sulfide

28. a. NH_4^+, ammonium ion
    b. SO_3^{2-}, sulfite ion
    c. NO_3^-, nitrate ion
    d. SO_4^{2-}, sulfate ion
    e. NO_2^-, nitrite ion
    f. CN^-, cyanide ion
    g. OH^-, hydroxide ion
    h. ClO_4^-, perchlorate ion
    i. ClO^-, hypochlorite ion
    j. PO_4^{3-}, phosphate ion

30. a. B_2O_3, diboron trioxide
    b. NO_2, nitrogen dioxide
    c. PCl_5, phosphorus trichloride
    d. N_2O_4, dinitrogen tetroxide
    e. P_2O_5, diphosphorus pentoxide
    f. ICl, iodine monochloride
    g. SF_6, sulfur hexafluoride
    h. N_2O_3, dinitrogen trioxide

Chapter 6   Chemical Reactions: An Introduction

2. The fact that there is a decrease in mass is the best evidence for reaction. If mass has been lost, then it is likely that a gaseous substance, which has escaped into the environment has been produced by the heating. The fact that the chalk crumbles into a powder may be taken as secondary evidence that the chalk has been converted into something which does not stick together well.

4. Hair certainly is not ordinarily soluble in water, yet when the depilatory is added, the hair dissolves and washes away.

6. The alcohol in the wine is converted by wild yeasts in the air into acetic acid (vinegar). The observation that the odor of the wine has changed to the odor of vinegar indicates that a new substance has been produced. This is a chemical reaction.

8. atoms

10. the same

12. water

14. $C_3H_8(g) + O_2(g) \rightarrow CO_2(g) + H_2O(g)$

16. $(NH_4)_2CO_3(s) \rightarrow NH_3(g) + CO_2(g) + H_2O(g)$

18. $CO(g) + H_2(g) \rightarrow CH_3OH(l)$

20. $Ca(s) + H_2O(l) \rightarrow Ca(OH)_2(s) + H_2(g)$

22. $Mg(OH)_2(s) + HCl(aq) \rightarrow MgCl_2(aq) + H_2O(l)$

24. $H_2S(g) + O_2(g) \rightarrow SO_2(g) + H_2O(g)$

26. $Fe_2O_3(s) + CO(g) \rightarrow Fe(l) + CO_2(g)$

28. $O_2(g) \rightarrow O_3(g)$

30. $NH_3(g) + HNO_3(aq) \rightarrow NH_4NO_3(s)$

32. $Xe(g) + F_2(g) \rightarrow XeF_4(s)$

34. $Ag(s) + HNO_3(aq) \rightarrow AgNO_3(aq) + H_2(g)$

36. whole numbers

38. a. $H_2O_2 \rightarrow H_2O + O_2$

   Balance oxygen: $\mathbf{2}H_2O_2 \rightarrow \mathbf{2}H_2O + O_2$

   Balanced equation: $\mathbf{2}H_2O_2(aq) \rightarrow \mathbf{2}H_2O(l) + O_2(g)$

Chapter 6   Chemical Reactions: An Introduction   51

    b. $Ag + H_2S \rightarrow Ag_2S + H_2$

        Balance silver: **2**$Ag + H_2S \rightarrow Ag_2S + H_2$

        Balanced equation: $2Ag(s) + H_2S(g) \rightarrow Ag_2S(s) + H_2(g)$

    c. $FeO + C \rightarrow Fe + CO_2$

        Balance oxygen: **2**$FeO + C \rightarrow Fe + CO_2$

        Balance iron: $2FeO + C \rightarrow$ **2**$Fe + CO_2$

        Balanced equation: $2FeO(s) + C(s) \rightarrow 2Fe(l) + CO_2(g)$

    d. $Cl_2 + KI \rightarrow KCl + I_2$

        Balance chlorine: $Cl_2 + KI \rightarrow$ **2**$KCl + I_2$

        Balance iodine: $Cl_2 +$ **2**$KI \rightarrow 2KCl + I_2$

        Balanced equation: $Cl_2(g) + 2KI(aq) \rightarrow 2KCl(aq) + I_2(s)$

    e. $Na_2B_4O_7 + H_2SO_4 + H_2O \rightarrow H_3BO_3 + Na_2SO_4$

        Balance boron: $Na_2B_4O_7 + H_2SO_4 + H_2O \rightarrow$ **4**$H_3BO_3 + Na_2SO_4$

        Balance hydrogen: $Na_2B_4O_7 + H_2SO_4 +$ **5**$H_2O \rightarrow 4H_3BO_3 + Na_2SO_4$

        Balanced equation: $Na_2B_4O_7(s) + H_2SO_4(aq) + 5H_2O(l)$
                                        $\rightarrow 4H_3BO_3(s) + Na_2SO_4(aq)$

    f. $CaC_2 + H_2O \rightarrow Ca(OH)_2 + C_2H_2$

        Balance oxygen: $CaC_2 +$ **2**$H_2O \rightarrow Ca(OH)_2 + C_2H_2$

        Balanced equation: $CaC_2(s) + 2H_2O(l) \rightarrow Ca(OH)_2(s) + C_2H_2(g)$

    g. $NaCl + H_2SO_4 \rightarrow HCl + Na_2SO_4$

        Balance sodium: **2**$NaCl + H_2SO_4 \rightarrow HCl + Na_2SO_4$

        Balance chlorine: $2NaCl + H_2SO_4 \rightarrow$ **2**$HCl + Na_2SO_4$

        Balanced equation: $2NaCl(s) + H_2SO_4(l) \rightarrow 2HCl(g) + Na_2SO_4(s)$

    h. $SiO_2 + C \rightarrow Si + CO$

        Balance oxygen: $SiO_2 + C \rightarrow Si +$ **2**CO

        Balance carbon: $SiO_2 +$ **2**$C \rightarrow Si + 2CO$

        Balanced equation: $SiO_2(s) + 2C(s) \rightarrow Si(l) + 2CO(g)$

40.   a. $CaF_2 + H_2SO_4 \rightarrow CaSO_4 + HF$

        Balance fluorine: $CaF_2 + H_2SO_4 \rightarrow CaSO_4 +$ **2**HF

        Balanced equation: $CaF_2(s) + H_2SO_4(l) \rightarrow CaSO_4(s) + 2HF(g)$

    b. $KBr + H_3PO_4 \rightarrow K_3PO_4 + HBr$

        Balance potassium: **3**$KBr + H_3PO_4 \rightarrow K_3PO_4 + HBr$

Balance bromine: $3KBr + H_3PO_4 \rightarrow K_3PO_4 + \mathbf{3}HBr$

Balanced equation: $3KBr(s) + H_3PO_4(aq) \rightarrow K_3PO_4(aq) + 3HBr(g)$

c. $TiCl_4 + Na \rightarrow NaCl + Ti$

Balance chlorine: $TiCl_4 + Na \rightarrow \mathbf{4}NaCl + Ti$

Balance sodium: $TiCl_4 + \mathbf{4}Na \rightarrow 4NaCl + Ti$

Balanced equation: $TiCl_4(l) + 4Na(s) \rightarrow 4NaCl(s) + Ti(s)$

d. $K_2CO_3 \rightarrow K_2O + CO_2$ This equation is already balanced!

e. $KO_2 + H_2O \rightarrow KOH + O_2$

Balance hydrogen: $KO_2 + H_2O \rightarrow \mathbf{2}KOH + O_2$

Balance potassium: $\mathbf{2}KO_2 + H_2O \rightarrow 2KOH + O_2$

At this point, we have balanced potassium and hydrogen atoms, but now it becomes difficult to balance oxygen since it occurs in each of the reactants and products. There is no systematic way to balance oxygen at this point (you will learn a special method in a later chapter for oxidation-reduction reactions such as this). We need one more oxygen atom on the right side of the equation to balance it: if we could just have an extra half of an O_2 molecule (that is, $1.5O_2$) the equation would be balanced. That is

Balance using non-integer: $2KO_2 + H_2O \rightarrow 2KOH + \mathbf{1.5}O_2$

Although this equation is balanced, we can't really have 1.5 molecules. If we multiply everything in this equation by 2, however, we will get whole number coefficients.

Balanced equation: $4KO_2(s) + 2H_2O(l) \rightarrow 4KOH(aq) + 3O_2(g)$

f. $Na_2O_2 + H_2O + CO_2 \rightarrow NaHCO_3 + O_2$

Balance sodium: $Na_2O_2 + H_2O + CO_2 \rightarrow \mathbf{2}NaHCO_3 + O_2$

Balance carbon: $Na_2O_2 + H_2O + \mathbf{2}CO_2 \rightarrow 2NaHCO_3 + O_2$

Again we are left with the difficulty of balancing oxygen when it occurs in all the reactants and products. Right now, there are 7 oxygen atoms on the left side of the equation, but there are 8 oxygen atoms on the right side. If we could have just half an O_2 molecule (instead of one entire molecule) things would work out.

Balance using non-integer: $Na_2O_2 + H_2O + 2CO_2 \rightarrow 2NaHCO_3 + \mathbf{0.5}O_2$

Although this equation is balanced, we can't really have half a molecule. If we multiply everything in the equation by 2, however, we'll get whole numbers.

Balanced: $2Na_2O_2(s) + 2H_2O(g) + 4CO_2(g) \rightarrow 4NaHCO_3(s) + O_2(g)$

g. $KNO_2 + C \rightarrow K_2CO_3 + CO + N_2$

Balance nitrogen: $\mathbf{2}KNO_2 + C \rightarrow K_2CO_3 + CO + N_2$

Balance carbon: $2KNO_2 + \mathbf{2}C \rightarrow K_2CO_3 + CO + N_2$

Balanced equation: $2KNO_2(s) + 2C(s) \rightarrow K_2CO_3(s) + CO(g) + N_2(g)$

h. $BaO + Al \rightarrow Ba + Al_2O_3$

Balance aluminum: $BaO + \mathbf{2}Al \rightarrow Ba + Al_2O_3$

Balance oxygen: $\mathbf{3}BaO + 2Al \rightarrow Ba + Al_2O_3$

Balance barium: $3BaO + 2Al \rightarrow \mathbf{3}Ba + Al_2O_3$

Balanced equation: $3BaO(s) + 2Al(s) \rightarrow 3Ba(s) + Al_2O_3(s)$

42. a. $SiI_4(s) + 2Mg(s) \rightarrow Si(s) + 2MgI_2(s)$
    b. $MnO_2(s) + 2Mg(s) \rightarrow Mn(s) + 2MgO(s)$
    c. $8Ba(s) + S_8(s) \rightarrow 8BaS(s)$
    d. $4NH_3(g) + 3Cl_2(g) \rightarrow 3NH_4Cl(s) + NCl_3(g)$
    e. $8Cu_2S(s) + S_8(s) \rightarrow 16CuS(s)$
    f. $2Al(s) + 3H_2SO_4(aq) \rightarrow Al_2(SO_4)_3(aq) + 3H_2(g)$
    g. $2NaCl(s) + H_2SO_4(l) \rightarrow 2HCl(g) + Na_2SO_4(s)$
    h. $2CO(g) + O_2(g) \rightarrow 2CO_2(g)$

44. a. $Ba(NO_3)_2(aq) + Na_2CrO_4(aq) \rightarrow BaCrO_4(s) + 2NaNO_3(aq)$
    b. $PbCl_2(aq) + K_2SO_4(aq) \rightarrow PbSO_4(s) + 2KCl(aq)$
    c. $C_2H_5OH(l) + 3O_2(g) \rightarrow 2CO_2(g) + 3H_2O(l)$
    d. $CaC_2(s) + 2H_2O(l) \rightarrow Ca(OH)_2(s) + C_2H_2(g)$
    e. $Sr(s) + 2HNO_3(aq) \rightarrow Sr(NO_3)_2(aq) + H_2(g)$
    f. $BaO_2(s) + H_2SO_4(aq) \rightarrow BaSO_4(s) + H_2O_2(aq)$
    g. $2AsI_3(s) \rightarrow 2As(s) + 3I_2(s)$
    h. $2CuSO_4(aq) + 4KI(s) \rightarrow 2CuI(s) + I_2(s) + 2K_2SO_4(aq)$

46. $Al(s) + O_2(g) \rightarrow Al_2O_3(s)$

48. $C_{12}H_{22}O_{11}(aq) + H_2O(l) \rightarrow 4C_2H_5OH(aq) + 4CO_2(g)$

50. $2Al_2O_3(s) + 3C(s) \rightarrow 4Al(s) + 3CO_2(g)$

52. $2Li(s) + S(s) \rightarrow Li_2S(s)$ $2Rb(s) + S(s) \rightarrow Rb_2S(s)$

    $2Na(s) + S(s) \rightarrow Na_2S(s)$ $2Cs(s) + S(s) \rightarrow Cs_2S(s)$

    $2K(s) + S(s) \rightarrow K_2S(s)$ $2Fr(s) + S(s) \rightarrow Fr_2S(s)$

54. $BaO_2(s) + H_2O(l) \rightarrow BaO(s) + H_2O_2(aq)$

56. $2KClO_3(s) \rightarrow 2KCl(s) + 3O_2(g)$

58. $NH_3(g) + HCl(g) \rightarrow NH_4Cl(s)$

60. The "charring" represents the conversion of the carbohydrates (starch) in the muffin to elemental carbon.

62. $Fe(s) + S(s) \rightarrow FeS(s)$

64. $K_2CrO_4(aq) + BaCl_2(aq) \rightarrow BaCrO_4(s) + 2KCl(aq)$

66. $2NaCl(aq) + 2H_2O(l) \rightarrow 2NaOH(aq) + H_2(g) + Cl_2(g)$

   $2NaBr(aq) + 2H_2O(l) \rightarrow 2NaOH(aq) + H_2(g) + Br_2(g)$

   $2NaI(aq) + 2H_2O(l) \rightarrow 2NaOH(aq) + H_2(g) + I_2(g)$

68. $CaC_2(s) + 2H_2O(l) \rightarrow Ca(OH)_2(s) + C_2H_2(g)$

70. $CuO(s) + H_2SO_4(aq) \rightarrow CuSO_4(aq) + H_2O(l)$

72. $Na_2SO_3(aq) + S(s) \rightarrow Na_2S_2O_3(aq)$

74. a. $ZnCl_2(aq) + Na_2CO_3(aq) \rightarrow ZnCO_3(s) + 2NaCl(aq)$
    b. $2Al(s) + 3H_2SO_4(aq) \rightarrow Al_2(SO_4)_3(aq) + 3H_2(g)$
    c. $Mn(s) + 2S(s) \rightarrow MnS_2(s)$
    d. $C_5H_{12}(l) + 8O_2(g) \rightarrow 5CO_2(g) + 6H_2O(g)$
    e. $H_2O(l) + Br_2(l) \rightarrow HBr(aq) + HOBr(aq)$
    f. $MnS_2(s) + 3O_2(g) \rightarrow MnO_2(s) + 2SO_2(g)$
    g. $PbCl_2(aq) + K_2CrO_4(aq) \rightarrow PbCrO_4(s) + 2KCl(aq)$
    h. $2AgNO_3(aq) + H_2SO_4(aq) \rightarrow Ag_2SO_4(s) + 2HNO_3(aq)$

76. a. $Pb(NO_3)_2(aq) + K_2CrO_4(aq) \rightarrow PbCrO_4(s) + 2KNO_3(aq)$
    b. $BaCl_2(aq) + Na_2SO_4(aq) \rightarrow BaSO_4(s) + 2NaCl(aq)$
    c. $2CH_3OH(l) + 3O_2(g) \rightarrow 2CO_2(g) + 4H_2O(g)$
    d. $Na_2CO_3(aq) + S(s) + SO_2(g) \rightarrow CO_2(g) + Na_2S_2O_3(aq)$
    e. $Cu(s) + 2H_2SO_4(aq) \rightarrow CuSO_4(aq) + SO_2(g) + 2H_2O(l)$
    f. $MnO_2(s) + 4HCl(aq) \rightarrow MnCl_2(aq) + Cl_2(g) + 2H_2O(l)$
    g. $As_2O_3(s) + 6KI(aq) + 6HCl(aq) \rightarrow 2AsI_3(s) + 6KCl(aq) + 3H_2O(l)$
    h. $2Na_2S_2O_3(aq) + I_2(aq) \rightarrow Na_2S_4O_6(aq) + 2NaI(aq)$

Chapter 7   Reactions in Aqueous Solutions

2. Driving forces are types of *changes* in a system which pull a reaction in the *direction of product formation*; driving forces discussed in Chapter Seven include: formation of a *solid*, formation of *water*, formation of a *gas*, transfer of electrons.

4. The net charge of a precipitate must be *zero*. The total number of positive charges equals the total number of negative charges.

6. ions

8. Chemists know that a solution contains separated ions because such a solution will readily allow an electrical current to pass through it. The simplest experiment that demonstrates this uses the sort of apparatus described in Figure 7.2: if the light bulb glows strongly, then the solution contains a strong electrolyte.

10. For most practical purposes, "insoluble" and "slightly" soluble mean the same thing. The difference between "insoluble" and "slightly soluble" could be crucial if, for example, a substance were highly toxic and were found in a water supply.

12. a. soluble (Rule 1: most nitrate salts are soluble)
    b. soluble (Rule 2: most potassium salts are soluble)
    c. soluble (Rule 2: most sodium salts are soluble)
    d. insoluble (Rule 5: most hydroxide compounds are insoluble)
    e. insoluble (Rule 3: exception for chloride salts)
    f. soluble (Rule 2: most ammonium salts are soluble)
    g. insoluble (Rule 6: most sulfide salts are insoluble)
    h. insoluble (Rule 4: exception for sulfate salts)

14. a. Rule 5: most hydroxides are only slightly soluble
    b. Rule 6: most carbonates are only slightly soluble
    c. Rule 6: most phosphates are only slightly soluble
    d. Rule 3: exception to the rule for chlorides

16. a. $CaSO_4$. Rule 4: exception to the rule for sulfates
    b. AgI. Rule 3: although the text does not mention it explicitly, as you might expect from your knowledge of the periodic table, bromide and iodide compounds of Ag^+, Pb^{2+}, and Hg_2^{2+} are insoluble
    c. $Pb_3(PO_4)_2$. Rule 6: most phosphate salts are only slightly soluble
    d. $Fe(OH)_3$. Rule 5: most hydroxides are only slightly soluble
    e. no precipitate is likely: rules 1, 2, and 4
    f. $BaCO_3$. Rule 6: most carbonate salts are only slightly soluble.

18. The precipitates are marked in boldface type.

   a. No precipitate: $Ba(NO_3)_2$ and HCl are each soluble.

   b. Rule 6: most sulfide salts are insoluble.
   $(NH_4)_2S(aq) + CoCl_2(aq) \rightarrow$ **CoS**$(s) + 2NH_4Cl(aq)$

   c. Rule 4: lead sulfate is a listed exception.
   $H_2SO_4(aq) + Pb(NO_3)_2(aq) \rightarrow$ **PbSO$_4$**$(s) + 2HNO_3(aq)$

   d. Rule 6: most carbonate salts are insoluble.
   $CaCl_2(aq) + K_2CO_3(aq) \rightarrow$ **CaCO$_3$**$(s) + 2KCl(aq)$

   e. No precipitate: $NaNO_3$ and $NH_4C_2H_3O_2$ are each soluble.

   f. Rule 6: most phosphate salts are insoluble
   $Na_3PO_4(aq) + CrCl_3(aq) \rightarrow 3NaCl(aq) +$ **CrPO$_4$**(s)

20. Hint: when balancing equations involving polyatomic ions, especially in precipitation reactions, balance the polyatomic ions as a *unit*, not in terms of the atoms the polyatomic ions contain (e.g., treat nitrate ion, NO_3^- as a single entity, not as one nitrogen and three oxygen atoms). When finished balancing, however, do be sure to count the individual number of atoms of each type on each side of the equation.

   a. $AgNO_3(aq) + H_2SO_4(aq) \rightarrow Ag_2SO_4(s) + HNO_3(aq)$
   Balance silver: **2**$AgNO_3(aq) + H_2SO_4(aq) \rightarrow Ag_2SO_4(s) + HNO_3(aq)$
   Balance nitrate: $2AgNO_3(aq) + H_2SO_4(aq) \rightarrow Ag_2SO_4(s) +$ **2**$HNO_3(aq)$
   Balanced equation: $2AgNO_3(aq) + H_2SO_4(aq) \rightarrow Ag_2SO_4(s) + 2HNO_3(aq)$

   b. $Ca(NO_3)_2(aq) + H_2SO_4(aq) \rightarrow CaSO_4(s) + HNO_3(aq)$
   Balance nitrate: $Ca(NO_3)_2(aq) + H_2SO_4(aq) \rightarrow CaSO_4(s) +$ **2**$HNO_3(aq)$
   Balanced equation: $Ca(NO_3)_2(aq) + H_2SO_4(aq) \rightarrow CaSO_4(s) + 2HNO_3(aq)$

   c. $Pb(NO_3)_2(aq) + H_2SO_4(aq) \rightarrow PbSO_4(s) + HNO_3(aq)$
   Balance nitrate: $Pb(NO_3)_2(aq) + H_2SO_4(aq) \rightarrow PbSO_4(s) +$ **2**$HNO_3(aq)$
   Balanced equation: $Pb(NO_3)_2(aq) + H_2SO_4(aq) \rightarrow PbSO_4(s) + 2HNO_3(aq)$

22. The products are determined by having the ions "switch partners." For example, for a general reaction $AB + CD \rightarrow$, the possible products are AD and CB if the ions switch partners. If either AD or CB is insoluble, then a precipitation reaction has occurred. In the following reaction, the formula of the precipitate is given in boldface type.

   a. $(NH_4)_2S(aq) + CoCl_2(aq) \rightarrow$ **CoS**$(s) + 2NH_4Cl(aq)$
   Rule 6: most sulfide salts are only slightly soluble

   b. $FeCl_3(aq) + 3NaOH(aq) \rightarrow$ **Fe(OH)$_3$**$(s) + 3NaCl(aq)$
   Rule 5: Most hydroxide compounds are only slightly soluble

Chapter 7   Reactions in Aqueous Solutions   57

    c. $CuSO_4(aq) + Na_2CO_3(aq) \rightarrow$ **$CuCO_3(s)$** $+ Na_2SO_4(aq)$

    Rule 6: most carbonate salts are only slightly soluble.

24. spectator

26. The net ionic equation for a reaction indicates *only those ions that go to form the precipitate*, and does not show the spectator ions present in the solutes mixed. The identity of the precipitate is determined from the Solubility Rules (Table 7.1).

    a. $Ca^{2+}(aq) + SO_4^{2-}(aq) \rightarrow CaSO_4(s)$

    Rule 4: exception to rule about sulfate salts.

    b. $2Fe^{3+}(aq) + 3CO_3^{2-}(aq) \rightarrow Fe_2(CO_3)_3(s)$

    Rule 6: most carbonate salts are only slightly soluble.

    c. $Ag^+(aq) + I^-(aq) \rightarrow AgI(s)$

    Rule 3: AgI, like AgCl, is insoluble.

    d. $3Co^{2+}(aq) + 3PO_4^{3-}(aq) \rightarrow Co_3(PO_4)_2(s)$

    Rule 6: most phosphate salts are only slightly soluble.

    e. $Hg_2^{2+}(aq) + 2Cl^-(aq) \rightarrow Hg_2Cl_2(s)$

    Rule 3: listed exception to the general rule about chlorides.

    f. $Pb^{2+}(aq) + 2Br^-(aq) \rightarrow PbBr_2(s)$

    Rule 3: like $PbCl_2$, $PbBr_2$ and PbI_2 are also insoluble.

28. $Ag^+(aq) + Cl^-(aq) \rightarrow AgCl(s)$

    $Pb^{2+}(aq) + 2Cl^-(aq) \rightarrow PbCl_2(s)$

    $Hg_2^{2+}(aq) + 2Cl^-(aq) \rightarrow Hg_2Cl_2(s)$

30. $Co^{2+}(aq) + S^{2-}(aq) \rightarrow CoS(s)$

    $2Co^{3+}(aq) + 3S^{2-}(aq) \rightarrow Co_2S_3(s)$

    $Fe^{2+}(aq) + S^{2-}(aq) \rightarrow FeS(s)$

    $2Fe^{3+}(aq) + 3S^{2-}(aq) \rightarrow Fe_2S_3(s)$

32. Strong bases are bases that fully produce hydroxide ions when dissolved in water. The strong bases are also strong electrolytes.

34. acids: HCl, H_2SO_4, HNO_3, $HClO_4$, HBr
    bases: NaOH, KOH, RbOH, CsOH

36. salt

58   Chapter 7   Reactions in Aqueous Solutions

38. RbOH(s) → Rb$^+$(aq) + OH$^-$(aq)
    CsOH(s) → Cs$^+$(aq) + OH$^-$(aq)

40. In general, the salt formed in an aqueous acid-base reaction consists of the *positive ion of the base* involved in the reaction, combined with the *negative ion of the acid*. The hydrogen ion of the strong acid combines with the hydroxide ion of the strong base to produce water, which is the other product of the acid-base reactions.

   a.   2NaOH(aq) + H$_2$SO$_4$(aq) → 2H$_2$O(l) + Na$_2$SO$_4$(aq)

   b.   RbOH(aq) + HNO$_3$(aq) → H$_2$O(l) + RbNO$_3$(aq)

   c.   KOH(aq) + HClO$_4$(aq) → H$_2$O(l) + KClO$_4$(aq)

   d.   KOH(aq) + HCl(aq) → H$_2$O(l) + KCl(aq)

42. transfer

44. The metallic element *loses* electrons and the nonmetallic element *gains* electrons.

46. Each nitrogen atom would gain three electrons to become an N^{3-} ion. Each nitrogen atom would gain three electrons, which means that an N$_2$ molecule would gain six electrons.

48. AlBr$_3$ is made up of Al^{3+} ions and Br$^-$ ions. Aluminum atoms each lose three electrons to become Al^{3+} ions. Bromine atoms each gain one electron to become Br$^-$ ions (so each Br$_2$ molecule gains two electrons to become two Br$^-$ ions).

50. a.   Fe(s) + S(s) → Fe$_2$S$_3$(s)

        Balance iron: **2**Fe + S → Fe$_2$S$_3$

        Balance sulfur: 2Fe + **3**S → Fe$_2$S$_3$

        Balanced equation: 2Fe(s) + 3S(s) → Fe$_2$S$_3$(s)

   b.   Zn(s) + HNO$_3$(aq) → Zn(NO$_3$)$_2$(aq) + H$_2$(g)

        Balance nitrate ions: Zn + **2**HNO$_3$ → Zn(NO$_3$)$_2$ + H$_2$

        Balanced equation: Zn(s) + 2HNO$_3$(aq) → Zn(NO$_3$)$_2$(aq) + H$_2$(g)

   c.   Sn(s) + O$_2$(g) → SnO(s)

        Balance oxygen: Sn + O$_2$ → **2**SnO

        Balance tin: **2**Sn + O$_2$ → 2SnO

        Balanced equation: 2Sn(s) + O$_2$(g) → 2SnO(s)

   d.   K(s) + H$_2$(g) → KH(s)

        Balance hydrogen: K + H$_2$ → **2**KH

        Balance potassium: **2**K + H$_2$ → 2KH

        Balanced equation: 2K(s) + H$_2$(g) → 2KH(s)

e. $Cs(s) + H_2O(l) \rightarrow CsOH(aq) + H_2(g)$

Balance hydrogen: $Cs + \mathbf{2}H_2O \rightarrow \mathbf{2}CsOH + H_2$

Balance cesium: $\mathbf{2}Cs + 2H_2O \rightarrow 2CsOH + H_2$

Balanced equation: $2Cs(s) + 2H_2O(l) \rightarrow 2CsOH(aq) + H_2(g)$

52. examples of formation of water:
$HCl(aq) + NaOH(aq) \rightarrow H_2O(l) + NaCl(aq)$
$H_2SO_4(aq) + 2KOH(aq) \rightarrow 2H_2O(l) + K_2SO_4(aq)$

examples of formation of a gaseous product
$Mg(s) + 2HCl(aq) \rightarrow MgCl_2(aq) + H_2(g)$
$2KClO_3(s) \rightarrow 2KCl(s) + 3O_2(g)$

54. For each reaction, the type of reaction is first identified, followed by some of the reasoning that leads to this choice (there may be more than one way in which you can recognize a particular type of reaction).

   a. oxidation-reduction (oxygen changes from the combined state to the elemental state)

   b. oxidation-reduction (copper changes from the elemental to the combined state; hydrogen changes from the combined to the elemental state)

   c. acid-base (H_2SO_4 is a strong acid and NaOH is a strong base; water and a salt are formed)

   d. acid-base, precipitation (H_2SO_4 is a strong acid, and $Ba(OH)_2$ is a base; water and a salt are formed; an insoluble product forms)

   e. precipitation (from the Solubility Rules of Table 7.1, AgCl is only slightly soluble)

   f. precipitation (from the Solubility Rules of Table 7.1, $Cu(OH)_2$ is only slightly soluble)

   g. oxidation-reduction (chlorine and fluorine change from the elemental to the combined state)

   h. oxidation-reduction (oxygen changes from the elemental to the combined state)

   i. acid-base (HNO_3 is a strong acid and $Ca(OH)_2$ is a strong base; a salt and water are formed)

56. oxidation-reduction

58. A decomposition reaction is one in which a given compound is broken down into simpler compounds or constituent elements. The reactions

$$CaCO_3(s) \to CaO(s) + CO_2(g)$$

$$2HgO(s) \to 2Hg(l) + O_2(g)$$

both represent decomposition reactions. Such reactions often (but not necessarily always) may be classified in other ways. For example, the reaction of HgO(s) is also an oxidation-reduction reaction.

60. Compounds like those in parts b and c of this problem, containing only carbon and hydrogen, are called *hydrocarbons*. When a hydrocarbon is reacted with oxygen (O_2), the hydrocarbon is almost always converted to carbon dioxide and water vapor. Since water molecules contain an odd number of oxygen atoms, whereas O_2 contains an even number of oxygen atoms, it is often difficult to balance such equations. For this reason, it is simpler to balance the equation using fractional coefficients if necessary, and then to multiply by a factor that will give whole number coefficients for the final balanced equation.

a. $C_2H_5OH(l) + O_2(g) \to CO_2(g) + H_2O(g)$

Balance carbon: $C_2H_5OH(l) + O_2(g) \to \mathbf{2}CO_2(g) + H_2O(g)$

Balance hydrogen: $C_2H_5OH(l) + O_2(g) \to 2CO_2(g) + \mathbf{3}H_2O(g)$

Balance oxygen: $C_2H_5OH(l) + \mathbf{3}O_2(g) \to 2CO_2(g) + 3H_2O(g)$

Balanced equation: $C_2H_5OH(l) + 3O_2(g) \to 2CO_2(g) + 3H_2O(g)$

b. $C_6H_{14}(l) + O_2(g) \to CO_2(g) + H_2O(g)$

Balance carbon: $C_6H_{14}(l) + O_2(g) \to \mathbf{6}CO_2(g) + H_2O(g)$

Balance hydrogen: $C_6H_{14}(l) + O_2(g) \to 6CO_2(g) + \mathbf{7}H_2O(g)$

Balance oxygen: $C_6H_{14}(l) + \mathbf{(19/2)}O_2(g) \to 6CO_2(g) + 7H_2O(g)$

Balanced equation: $2C_6H_{14}(l) + 19O_2(g) \to 12CO_2(g) + 14H_2O(g)$

c. $C_6H_{12}(l) + O_2(g) \to CO_2(g) + H_2O(g)$

Balance carbon: $C_6H_{12}(l) + O_2(g) \to \mathbf{6}CO_2(g) + H_2O(g)$

Balance hydrogen: $C_6H_{12}(l) + O_2(g) \to 6CO_2(g) + \mathbf{6}H_2O(g)$

Balanced equation: $C_6H_{12}(l) + 9O_2(g) \to 6CO_2(g) + 6H_2O(g)$

62. a. $C_2H_6(g) + O_2(g) \to CO_2(g) + H_2O(g)$

Balance carbon: $C_2H_6(g) + O_2(g) \to \mathbf{2}CO_2(g) + H_2O(g)$

Balance hydrogen: $C_2H_6(g) + O_2(g) \to 2CO_2(g) + \mathbf{3}H_2O(g)$

Balance oxygen: $C_2H_6(g) + \mathbf{(7/2)}O_2(g) \to 2CO_2(g) + 3H_2O(g)$

Balanced equation: $2C_2H_6(g) + 7O_2(g) \to 4CO_2(g) + 6H_2O(g)$

b. $C_2H_6O(l) + O_2(g) \rightarrow CO_2(g) + H_2O(g)$

Balance carbon: $C_2H_6O(l) + O_2(g) \rightarrow \mathbf{2}CO_2(g) + H_2O(g)$

Balance hydrogen: $C_2H_6O(l) + O_2(g) \rightarrow 2CO_2(g) + \mathbf{3}H_2O(g)$

Balance oxygen: $C_2H_6O(l) + \mathbf{3}O_2(g) \rightarrow 2CO_2(g) + 3H_2O(g)$

Balanced equation: $C_2H_6O(l) + 3O_2(g) \rightarrow 2CO_2(g) + 3H_2O(g)$

c. $C_2H_6O_2(l) + O_2(g) \rightarrow CO_2(g) + H_2O(g)$

Balance carbon: $C_2H_6O_2(l) + O_2(g) \rightarrow \mathbf{2}CO_2(g) + H_2O(g)$

Balance hydrogen: $C_2H_6O_2(l) + O_2(g) \rightarrow 2CO_2(g) + \mathbf{3}H_2O(g)$

Balance oxygen: $C_2H_6O_2(l) + \mathbf{(5/2)}O_2(g) \rightarrow 2CO_2(g) + 3H_2O(g)$

Balanced equation: $2C_2H_6O_2(l) + 5O_2(g) \rightarrow 4CO_2(g) + 6H_2O(g)$

64. a. $2Co(s) + 3S(s) \rightarrow Co_2S_3(s)$

b. $2NO(g) + O_2(g) \rightarrow 2NO_2(g)$

c. $FeO(s) + CO_2(g) \rightarrow FeCO_3(s)$

d. $2Al(s) + 3F_2(g) \rightarrow 2AlF_3(s)$

e. $2NH_3(g) + H_2CO_3(aq) \rightarrow (NH_4)_2CO_3(s)$

66. a. $2NI_3(s) \rightarrow N_2(g) + 3I_2(s)$

b. $BaCO_3(s) \rightarrow BaO(s) + CO_2(g)$

c. $C_6H_{12}O_6(s) \rightarrow 6C(s) + 6H_2O(g)$

d. $Cu(NH_3)_4SO_4(s) \rightarrow CuSO_4(s) + 4NH_3(g)$

e. $3NaN_3(s) \rightarrow Na_3N(s) + 4N_2(g)$

68. In several cases, the given ion may be precipitated by *many* reactants. The following are only three of the possible examples.

a. chloride ion would precipitate when treated with solutions containing silver ion, lead(II) ion, or mercury(I) ion.

$Ag^+(aq) + Cl^-(aq) \rightarrow AgCl(s)$

$Pb^{2+}(aq) + 2Cl^-(aq) \rightarrow PbCl_2(s)$

$Hg_2^{2+}(aq) + 2Cl^-(aq) \rightarrow Hg_2Cl_2(s)$

b. calcium ion would precipitate when treated with solutions containing sulfate ion, carbonate ion, and phosphate ion.

$Ca^{2+}(aq) + SO_4^{2-}(aq) \rightarrow CaSO_4(s)$

$Ca^{2+}(aq) + CO_3^{2-}(aq) \rightarrow CaCO_3(s)$

$3Ca^{2+}(aq) + 2PO_4^{3-}(aq) \rightarrow Ca_3(PO_4)_2(s)$

c. iron(III) ion would precipitate when treated with solutions containing hydroxide, sulfide, or carbonate ions.

$Fe^{3+}(aq) + 3OH^-(aq) \rightarrow Fe(OH)_3(s)$

$2Fe^{3+}(aq) + 3S^{2-}(aq) \rightarrow Fe_2S_3(s)$

$2Fe^{3+}(aq) + 3CO_3^{2-}(aq) \rightarrow Fe_2(CO_3)_3(s)$

d. sulfate ion would precipitate when treated with solutions containing barium ion, calcium ion, or lead(II) ion.

$Ba^{2+}(aq) + SO_4^{2-}(aq) \rightarrow BaSO_4(s)$

$Ca^{2+}(aq) + SO_4^{2-}(aq) \rightarrow CaSO_4(s)$

$Pb^{2+}(aq) + SO_4^{2-}(aq) \rightarrow PbSO_4(s)$

e. mercury(I) ion would precipitate when treated with solutions containing chloride ion, sulfide ion, or carbonate ion.

$Hg_2^{2+}(aq) + 2Cl^-(aq) \rightarrow Hg_2Cl_2(s)$

$Hg_2^{2+}(aq) + S^{2-}(aq) \rightarrow Hg_2S(s)$

$Hg_2^{2+}(aq) + CO_3^{2-}(aq) \rightarrow Hg_2CO_3(s)$

f. silver ion would precipitate when treated with solutions containing chloride ion, sulfide ion, or carbonate ion.

$Ag^+(aq) + Cl^-(aq) \rightarrow AgCl(s)$

$2Ag^+(aq) + S^{2-}(aq) \rightarrow Ag_2S(s)$

$2Ag^+(aq) + CO_3^{2-}(aq) \rightarrow Ag_2CO_3(s)$

70. The formulas of the salts are indicated in boldface type.

a. $HNO_3(aq) + KOH(aq) \rightarrow H_2O(l) +$ **KNO_3**(aq)

b. $H_2SO_4(aq) + Ba(OH)_2(aq) \rightarrow 2H_2O(l) +$ **$BaSO_4$**(s)

c. $HClO_4(aq) + NaOH(aq) \rightarrow H_2O(l) +$ **$NaClO_4$**(aq)

d. $2HCl(aq) + Ca(OH)_2(aq) \rightarrow 2H_2O(l) +$ **$CaCl_2$**(aq)

72. a. soluble (Rule 2: most potassium salts are soluble)

b. soluble (Rule 2: most ammonium salts are soluble)

c. insoluble (Rule 6: most carbonate salts are only slightly soluble)

d. insoluble (Rule 6: most phosphate salts are only slightly soluble)

e. soluble (Rule 2: most sodium salts are soluble)

f. insoluble (Rule 6: most carbonate salts are only slightly soluble)

g. soluble (Rule 3: most chloride salts are soluble)

Chapter 7 Reactions in Aqueous Solutions

74. The precipitates are marked in boldface type.

   a. Rule 3: AgCl is listed as an exception

   $AgNO_3(aq) + HCl(aq) \rightarrow$ **AgCl**$(s) + HNO_3(aq)$

   b. Rule 6: most cabonate salts are only slightly soluble

   $CuSO_4(aq) + (NH_4)_2CO_3(aq) \rightarrow$ **CuCO**$_3(s) + (NH_4)_2SO_4(aq)$

   c. Rule 6: most carbonate salts are only slightly soluble.

   $FeSO_4(aq) + K_2CO_3(aq) \rightarrow$ **FeCO**$_3(s) + K_2SO_4(aq)$

   d. no reaction

   e. Rule 6: most carbonate salts are only slightly soluble

   $Pb(NO_3)_2(aq) + Li_2CO_3(aq) \rightarrow$ **PbCO**$_3(s) + 2LiNO_3(aq)$

   f. Rule 5: most hydroxide compounds are only slightly soluble

   $SnCl_4(aq) + 4NaOH(aq) \rightarrow$ **Sn(OH)**$_4(s) + 4NaCl(aq)$

76. $Fe^{2+}(aq) + S^{2-}(aq) \rightarrow FeS(s)$

   $2Cr^{3+}(aq) + 3S^{2-}(aq) \rightarrow Cr_2S_3(s)$

   $Ni^{2+}(aq) + S^{2-}(aq) \rightarrow NiS(s)$

78. These anions tend to form insoluble precipitates with *many* metal ions. The following are illustrative for cobalt(II) chloride, tin(II) chloride, and copper(II) nitrate reacting with the sodium salts of the given anions.

   a. $CoCl_2(aq) + Na_2S(aq) \rightarrow CoS(s) + 2NaCl(aq)$

   $SnCl_2(aq) + Na_2S(aq) \rightarrow SnS(s) + 2NaCl(aq)$

   $Cu(NO_3)_2(aq) + Na_2S(aq) \rightarrow CuS(s) + 2NaNO_3(aq)$

   b. $CoCl_2(aq) + Na_2CO_3(aq) \rightarrow CoCO_3(s) + 2NaCl(aq)$

   $SnCl_2(aq) + Na_2CO_3(aq) \rightarrow SnCO_3(s) + 2NaCl(aq)$

   $Cu(NO_3)_2(aq) + Na_2CO_3(aq) \rightarrow CuCO_3(s) + 2NaNO_3(aq)$

   c. $CoCl_2(aq) + 2NaOH(aq) \rightarrow Co(OH)_2(s) + 2NaCl(aq)$

   $SnCl_2(aq) + 2NaOH(aq) \rightarrow Sn(OH)_2(s) + 2NaCl(aq)$

   $Cu(NO_3)_2(aq) + 2NaOH(aq) \rightarrow Cu(OH)_2(s) + 2NaNO_3(aq)$

   d. $3CoCl_2(aq) + 2Na_3PO_4(aq) \rightarrow Co_3(PO_4)_2(s) + 6NaCl(aq)$

   $3SnCl_2(aq) + 2Na_3PO_4(aq) \rightarrow Sn_3(PO_4)_2(s) + 6NaCl(aq)$

   $3Cu(NO_3)_2(aq) + 2Na_3PO_4(aq) \rightarrow Cu_3(PO_4)_2(s) + 6NaNO_3(aq)$

80. a. $Na + O_2 \rightarrow Na_2O_2$
   Balance sodium: $2Na + O_2 \rightarrow Na_2O_2$
   Balanced equation: $2Na(s) + O_2(g) \rightarrow Na_2O_2(s)$

   b. $Fe(s) + H_2SO_4(aq) \rightarrow FeSO_4(aq) + H_2(g)$
   Equation is already balanced!

   c. $Al_2O_3 \rightarrow Al + O_2$
   Balance oxygen: $2Al_2O_3 \rightarrow Al + 3O_2$
   Balance aluminum: $2Al_2O_3 \rightarrow 4Al + 3O_2$
   Balanced equation: $2Al_2O_3(s) \rightarrow 4Al(s) + 3O_2(g)$

   d. $Fe + Br_2 \rightarrow FeBr_3$
   Balance bromine: $Fe + 3Br_2 \rightarrow 2FeBr_3$
   Balance iron: $2Fe + 3Br_2 \rightarrow 2FeBr_3$
   Balanced equation: $2Fe(s) + 3Br_2(l) \rightarrow 2FeBr_3(s)$

   e. $Zn + HNO_3 \rightarrow Zn(NO_3)_2 + H_2$
   Balance nitrate ions: $Zn + 2HNO_3 \rightarrow Zn(NO_3)_2 + H_2$
   Balanced equation: $Zn(s) + 2HNO_3(aq) \rightarrow Zn(NO_3)_2(aq) + H_2(g)$

82. a. $2C_4H_{10}(l) + 13O_2(g) \rightarrow 8CO_2(g) + 10H_2O(g)$
   b. $C_4H_{10}O(l) + 6O_2(g) \rightarrow 4CO_2(g) + 5H_2O(g)$
   c. $2C_4H_{10}O_2(l) + 11O_2(g) \rightarrow 8CO_2(g) + 10H_2O(g)$

84. a. $2NaHCO_3(s) \rightarrow Na_2CO_3(s) + H_2O(g) + CO_2(g)$
   b. $2NaClO_3(s) \rightarrow 2NaCl(s) + 3O_2(g)$
   c. $2HgO(s) \rightarrow 2Hg(l) + O_2(g)$
   d. $C_{12}H_{22}O_{11}(s) \rightarrow 12C(s) + 11H_2O(g)$
   e. $2H_2O_2(l) \rightarrow 2H_2O(l) + O_2(g)$

86. $Fe(s) + H_2SO_4(aq) \rightarrow FeSO_4(aq) + H_2(g)$
   $Zn(s) + H_2SO_4(aq) \rightarrow ZnSO_4(aq) + H_2(g)$
   $Mg(s) + H_2SO_4(aq) \rightarrow MgSO_4(aq) + H_2(g)$
   $Co(s) + H_2SO_4(aq) \rightarrow CoSO_4(aq) + H_2(g)$
   $Ni(s) + H_2SO_4(aq) \rightarrow NiSO_4(aq) + H_2(g)$

88.
a. one
b. one
c. two
d. two
e. three

90. A very simple example which fits the bill is: $C(s) + O_2(g) \rightarrow CO_2(g)$

92.
a. $2C_3H_8O(l) + 9O_2(g) \rightarrow 6CO_2(g) + 8H_2O(g)$
oxidation-reduction, combustion

b. $HCl(aq) + AgC_2H_3O_2(aq) \rightarrow AgCl(s) + HC_2H_3O_2(aq)$
precipitation, double-displacement

c. $3HCl(aq) + Al(OH)_3(s) \rightarrow AlCl_3(aq) + 3H_2O(l)$
acid-base, double-displacement

d. $2H_2O_2(aq) \rightarrow 2H_2O(l) + O_2(g)$
oxidation-reduction, decomposition

e. $N_2H_4(l) + O_2(g) \rightarrow N_2(g) + 2H_2O(g)$
oxidation-reduction, combustion

94. $2Na(s) + Cl_2(g) \rightarrow 2NaCl(s)$

$2Al(s) + 3Cl_2(g) \rightarrow 2AlCl_3(s)$

$Zn(s) + Cl_2(g) \rightarrow ZnCl_2(s)$

$Ca(s) + Cl_2(g) \rightarrow CaCl_2(s)$

$2Fe(s) + 3Cl_2(g) \rightarrow 2FeCl_3(s)$; $Fe(s) + Cl_2(g) \rightarrow FeCl_2(s)$

Cumulative Review: Chapters 6 and 7

2. A chemical equation indicates the substances necessary for a chemical reaction to take place, as well as what is produced by that chemical reaction. The substances to the left of the arrow in a chemical equation are called the reactants; those to the right of the arrow are referred to as the products. In addition, if a chemical equation has been balanced, then the equation indicates the relative proportions in which the reactant molecules combine to form the product molecules.

4. It is *never* permissible to change the subscripts of a formula when balancing a chemical equation: changing the subscripts changes the *identity* of a substance from one chemical to another. For example, consider the unbalanced chemical equation

    $$H_2(g) + O_2(g) \rightarrow H_2O(l)$$

    If you changed the *formula* of the product from $H_2O(l)$ to $H_2O_2(l)$, the equation would appear to be "balanced". However, H_2O is water, whereas H_2O_2 is hydrogen peroxide--a completely different chemical substance (which is not prepared by reaction of the elements hydrogen and oxygen).
    When we balance a chemical equation, it is permitted only to adjust the *coefficients* of a formula, since changing a coefficient merely changes the number of molecules of a substance being used in the reaction, without changing the identity of the substance. For the example above, we can balance the equation by putting coefficients of 2 in front of the formulas of H_2 and H_2O: these coefficients do not change the nature of what is reacting and what product is formed.

    $$2H_2(g) + O_2(g) \rightarrow 2H_2O(l)$$

6. A precipitation reaction is one in which a *solid* forms when the reactants are combined: the solid is called a precipitate. If you were to perform such a reaction, the mixture would turn cloudy as the reactants are combined, and a solid would eventually settle from the mixture on standing. There are many examples of such precipitation reactions: consult the solubility rules in Table 7.1 if you need help. One example would be to combine barium nitrate and sodium carbonate solutions: a precipitate of barium carbonate would form.

    $$Ba(NO_3)_2(aq) + Na_2CO_3(aq) \rightarrow BaCO_3(s) + 2NaNO_3(aq)$$

8. In summary, nearly all compounds containing the nitrate, sodium, potassium, and ammonium ions are soluble in water. Most salts containing the chloride and sulfate ions are soluble in water, with specific exceptions (see Table 7.1 for these exceptions). Most compounds containing the hydroxide, sulfide, carbonate, and phosphate ions are not soluble in water, unless the compound also contains one of the cations mentioned above (Na^+, K^+, NH_4^+).
    The solubility rules are phrased as if you had a sample of a given solute and wanted to see if you could dissolve it in water. These rules can also be applied, however, to predict the identity of the solid produced in a precipitation reaction: a given combination of ions will not be soluble in water whether you take a pure compound out of a reagent bottle or if you generate the insoluble combination of ions during a

chemical reaction. For example, the solubility rules say that $BaSO_4$ is not soluble in water. This means not only that a pure sample of $BaSO_4$ taken from a reagent bottle will not dissolve in water, but also that if Ba^{2+} ion and SO_4^{2-} ion end up together in the same solution, they will precipitate as $BaSO_4$. If we were to combine barium chloride and sulfuric acid solutions

$$BaCl_2(aq) + H_2SO_4(aq) \rightarrow BaSO_4(s) + 2HCl(aq)$$

then, because barium sulfate is not soluble in water, a precipitate of $BaSO_4(s)$ would form. Since a precipitate of $BaSO_4(s)$ would form no matter what barium compound or what sulfate compound were mixed, we can write the net ionic equation for the reaction as

$$Ba^{2+}(aq) + SO_4^{2-}(aq) \rightarrow BaSO_4(s)$$

Thus if, for example, barium nitrate solution were combined with sodium sulfate solution, a precipitate of $BaSO_4$ would form. Barium sulfate is insoluble in water regardless of its source.

10. Acids (such as the citric acid found in citrus fruits and the acetic acid found in vinegar) were first noted primarily because of their sour taste. The first bases noted were characterized by their bitter taste and slippery feel on the skin. Acids and bases chemically react with (neutralize) each other forming water: the net ionic equation is

$$H^+(aq) + OH^-(aq) \rightarrow H_2O(l)$$

The *strong* acids and bases are those which fully ionize when they dissolve in water: since these substances fully ionize, they are strong electrolytes. The common strong acids are HCl(hydrochloric), HNO_3(nitric), H_2SO_4(sulfuric), and $HClO_4$(perchloric). The most common strong bases are the alkali metal hydroxides, particularly NaOH(sodium hydroxide) and KOH(potassium hydroxide).

12. Oxidation-reduction reactions are electron-transfer reactions. Oxidation represents a loss of electrons by an atom, molecule, or ion, whereas reduction is the gain of electrons by such a species. Since an oxidation-reduction process represents the transfer of electrons between species, you can't have one without the other also taking place: the electrons lost by one species must be gained by some other species. An example of a simple oxidation reduction reaction between a metal and a nonmetal could be the following

$$Mg(s) + F_2(g) \rightarrow MgF_2(s)$$

In this process, Mg atoms lose two electrons each to become Mg^{2+} ions in MgF_2: Mg is oxidized. Each F atom of F_2 gains one electron to become an F^- ion, for a total of two electrons gained for each F_2 molecule: F_2 is reduced.

$$Mg \rightarrow Mg^{2+} + 2e^- \qquad 2(F + e^- \rightarrow F^-)$$

68   Cumulative Review: Chapters 6 and 7

14.   In general, a synthesis reaction represents the reaction of elements or simple compounds to produce more complex substances. There are many examples of synthesis reactions, for example

$$N_2(g) + 3H_2(g) \rightarrow 2NH_3(g)$$

$$NaOH(aq) + CO_2(g) \rightarrow NaHCO_3(s)$$

Decomposition reactions represent the breakdown of a more complex substance into simpler substances. There are many examples of decomposition reactions, for example

$$2H_2O_2(aq) \rightarrow 2H_2O(l) + O_2(g)$$

Synthesis and decomposition reactions are very often also oxidation-reduction reactions, especially if an elemental substance reacts or is generated. It is not necessary, however, for synthesis and decomposition reactions to always involve oxidation-reduction. For example, the reaction between NaOH and CO_2 given as an example of a synthesis reaction does *not* represent oxidation-reduction.

16.   a. $2Na(s) + 2H_2O(l) \rightarrow 2NaOH(aq) + H_2(g)$

           $2K(s) + 2H_2O(l) \rightarrow 2KOH(aq) + H_2(g)$

      b. $2Na(s) + Cl_2(g) \rightarrow 2NaCl(s)$

           $2K(s) + Cl_2(g) \rightarrow 2KCl(s)$

      c. $3Na(s) + P(s) \rightarrow Na_3P(s)$

           $3K(s) + P(s) \rightarrow K_3P(s)$

      d. $6Na(s) + N_2(g) \rightarrow 2Na_3N(s)$

           $6K(s) + N_2(g) \rightarrow 2K_3N(s)$

      e. $2Na(s) + H_2(g) \rightarrow 2NaH(s)$

           $2K(s) + H_2(g) \rightarrow 2KH(s)$

18.   a. $Ba(NO_3)_2(aq) + K_2CrO_4(aq) \rightarrow BaCrO_4(s) + 2KNO_3(aq)$

      b. $NaOH(aq) + CH_3COOH(aq) \rightarrow H_2O(l) + NaCH_3COO(aq)$
           (then evaporate the water from the solution)

      c. $AgNO_3(aq) + NaCl(aq) \rightarrow AgCl(s) + NaNO_3(aq)$

      d. $Pb(NO_3)_2(aq) + H_2SO_4(aq) \rightarrow PbSO_4(s) + 2HNO_3(aq)$

      e. $2NaOH(aq) + H_2SO_4(aq) \rightarrow Na_2SO_4(aq) + 2H_2O(l)$
           (then evaporate the water from the solution)

      f. $Ba(NO_3)_2(aq) + Na_2CO_3(aq) \rightarrow BaCO_3(s) + 2NaNO_3(aq)$

20. a. $FeO(s) + 2HNO_3(aq) \rightarrow Fe(NO_3)_2(aq) + H_2O(l)$
acid-base, double-displacement

b. $2Mg(s) + 2CO_2(g) + O_2(g) \rightarrow 2MgCO_3(s)$
synthesis; oxidation-reduction

c. $2NaOH(s) + CuSO_4(aq) \rightarrow Cu(OH)_2(s) + Na_2SO_4(aq)$
precipitation, double-displacement

d. $HI(aq) + KOH(aq) \rightarrow KI(aq) + H_2O(l)$
acid-base, double-displacement

e. $C_3H_8(g) + 5O_2(g) \rightarrow 3CO_2(g) + 4H_2O(g)$
combustion; oxidation-reduction

f. $Co(NH_3)_6Cl_2(s) \rightarrow CoCl_2(s) + 6NH_3(g)$
decomposition

g. $2HCl(aq) + Pb(C_2H_3O_2)_2(aq) \rightarrow 2HC_2H_3O_2(aq) + PbCl_2(s)$
precipitation, double-displacement

h. $C_{12}H_{22}O_{11}(s) \rightarrow 12C(s) + 11H_2O(g)$
decomposition; oxidation-reduction

i. $2Al(s) + 6HNO_3(aq) \rightarrow 2Al(NO_3)_3(aq) + 3H_2(g)$
oxidation-reduction; single-displacement

j. $4B(s) + 3O_2(g) \rightarrow 2B_2O_3(s)$
synthesis; oxidation-reduction

Chapter 8   Chemical Composition

2. $500.\ g \times \dfrac{1\ cork}{1.63\ g} = 306.7 = 307\ corks$

   $500.\ g \times \dfrac{1\ stopper}{4.31\ g} = 116\ stoppers$

   1 kg (1000 g) of corks contains $(1000\ g \times \dfrac{1\ cork}{1.63\ g}) = 613.49 = 613\ corks$

   613 stoppers would weigh $(613\ stoppers \times \dfrac{4.31\ g}{1\ stopper}) = 2644\ g = 2640\ g$

   The ratio of the mass of a stopper to the mass of a cork is (4.31 g/1.63 g). So the mass of stoppers that contains the same number of stoppers as there are corks in 1000 g of corks is

   $1000\ g \times \dfrac{4.31\ g}{1.63\ g} = 2644\ g = 2640\ g$

4. We use the average atomic mass of an element when performing calculations because the average mass takes into account the individual masses and relative abundances of all the isotopes of an element.

6. a. $10.81\ amu \times \dfrac{1\ B\ atom}{10.81\ amu} = 1.000\ atom = 1\ B\ atom$

    b. $320.7\ amu \times \dfrac{1\ S\ atom}{32.07\ amu} = 10\ S\ atoms$

    c. $19{,}691\ amu \times \dfrac{1\ Au\ atom}{197.97\ amu} = 100.00\ Au\ atoms = 100\ Au\ atoms$

    d. $19{,}695\ amu \times \dfrac{1\ Xe\ atom}{131.3\ amu} = 150.0\ Xe\ atoms = 150\ Xe\ atoms$

    e. $3588.3\ amu \times \dfrac{1\ Al\ atom}{26.98\ amu} = 133.0\ Al\ atoms = 133\ Al\ atoms$

8. $1.00 \times 10^4\ atoms \times \dfrac{196.97\ amu}{1\ Au\ atom} = 1.97 \times 10^6\ amu$

   $2.955 \times 10^5\ amu \times \dfrac{1\ Au\ atom}{196.97\ amu} = 1500.\ Au\ atoms$

10. Avogadro's number (6.022×10^{23})

12. The ratio of the atomic mass of Fe to the atomic mass of N is (55.85 amu/14.01 amu), and the mass of iron is given by

   $14.01\ g \times \dfrac{55.85\ amu}{14.01\ amu} = 55.85\ g\ Fe$

14. The ratio of the atomic mass of Co to the atomic mass of F is (58.93 amu/19.00 amu), and the mass of cobalt is given by

$$57.0 \text{ g} \times \frac{58.93 \text{ amu}}{19.00 \text{ amu}} = 177 \text{ g Co}$$

16. 1 mol O = 16.00 g O = 6.02×10^{23} O atoms

$$1 \text{ O atom} \times \frac{16.00 \text{ g O}}{6.022 \times 10^{23} \text{ O atoms}} = 2.66 \times 10^{-23} \text{ g}$$

18. $0.50 \text{ mol O atoms} \times \dfrac{16.00 \text{ g}}{1 \text{ mol}} = 8.0 \text{ g O}$

$$4 \text{ mol H atoms} \times \frac{1.008 \text{ g}}{1 \text{ mol}} = 4 \text{ g H}$$

Half a mole of O atoms weighs more than 4 moles of H atoms.

20.
 a. $26.2 \text{ g Au} \times \dfrac{1 \text{ mol Au}}{197.0 \text{ g}} = 0.133 \text{ mol Au}$

 b. $41.5 \text{ g Ca} \times \dfrac{1 \text{ mol Ca}}{40.08 \text{ g}} = 1.04 \text{ mol Ca}$

 c. $335 \text{ mg Ba} \times \dfrac{1 \text{ g}}{10^3 \text{ mg}} \times \dfrac{1 \text{ mol Ba}}{137.3 \text{ g}} = 2.44 \times 10^{-3} \text{ mol Ba}$

 d. $1.42 \times 10^{-3} \text{ g Pd} \times \dfrac{1 \text{ mol Pd}}{106.4 \text{ g}} = 1.33 \times 10^{-5} \text{ mol Pd}$

 e. $3.05 \times 10^{-5} \text{ μg Ni} \times \dfrac{1 \text{ g}}{10^6 \text{ μg}} \times \dfrac{1 \text{ mol Ni}}{58.70 \text{ g}} = 5.20 \times 10^{-13} \text{ mol Ni}$

 f. $1.00 \text{ lb Fe} \times \dfrac{453.59 \text{ g}}{1 \text{ lb}} \times \dfrac{1 \text{ mol Fe}}{55.85 \text{ g}} = 8.12 \text{ mol Fe}$

 g. $12.01 \text{ g C} \times \dfrac{1 \text{ mol C}}{12.01 \text{ g}} = 1.000 \text{ mol C}$

22.
 a. $2.00 \text{ mol Fe} \times \dfrac{55.85 \text{ g}}{1 \text{ mol}} = 112 \text{ g Fe}$

 b. $0.521 \text{ mol Ni} \times \dfrac{58.70 \text{ g}}{1 \text{ mol}} = 30.6 \text{ g Ni}$

 c. $1.23 \times 10^{-3} \text{ mol Pt} \times \dfrac{195.1 \text{ g}}{1 \text{ mol}} = 0.240 \text{ g Pt}$

72   Chapter 8   Chemical Composition

d. $72.5 \text{ mol Pb} \times \dfrac{207.2 \text{ g}}{1 \text{ mol}} = 1.50 \times 10^4 \text{ g Pb}$

e. $0.00102 \text{ mol Mg} \times \dfrac{24.31 \text{ g}}{1 \text{ mol}} = 0.0248 \text{ g Mg}$

f. $4.87 \times 10^3 \text{ mol Al} \times \dfrac{26.98 \text{ g}}{1 \text{ mol}} = 1.31 \times 10^5 \text{ g Al}$

g. $211.5 \text{ mol Li} \times \dfrac{6.941 \text{ g}}{1 \text{ mol}} = 1468 \text{ g Li}$

h. $1.72 \times 10^{-6} \text{ mol Na} \times \dfrac{22.99 \text{ g}}{1 \text{ mol}} = 3.95 \times 10^{-5} \text{ g Na}$

24. a. $0.00103 \text{ g Co} \times \dfrac{6.022 \times 10^{23} \text{ Co atoms}}{58.93 \text{ g Co}} = 1.05 \times 10^{19} \text{ Co atoms}$

    b. $0.00103 \text{ mol Co} \times \dfrac{6.022 \times 10^{23} \text{ Co atoms}}{1 \text{ mol}} = 6.20 \times 10^{20} \text{ Co atoms}$

    c. $2.75 \text{ g cobalt} \times \dfrac{1 \text{ mol}}{58.93 \text{ g Co}} = 0.0467 \text{ mol Co}$

    d. $5.99 \times 10^{21} \text{ Co atoms} \times \dfrac{1 \text{ mol}}{6.022 \times 10^{23} \text{ Co atoms}} = 0.00995 \text{ mol Co}$

    e. $4.23 \text{ mol Co} \times \dfrac{58.93 \text{ g Co}}{1 \text{ mol Co}} = 249 \text{ g Co}$

    f. $4.23 \text{ mol Co} \times \dfrac{6.022 \times 10^{23} \text{ Co atoms}}{1 \text{ mol Co}} = 2.55 \times 10^{24} \text{ Co atoms}$

    g. $4.23 \text{ g Co} \times \dfrac{6.022 \times 10^{23} \text{ Co atoms}}{58.93 \text{ g Co}} = 4.32 \times 10^{22} \text{ Co atoms}$

26. adding together (summing)

28. a. mass of 3 mol Na = 3(22.99 g) = 68.97 g
       mass of 1 mol N = 1(14.01 g) = 14.01 g
       molar mass of Na_3N =   82.98 g

    b. mass of 1 mol C =   12.01 g = 12.01 g
       mass of 2 mol S = 2(32.07 g) = 64.14 g
       molar mass of CS_2 =   76.15 g

    c. mass of 1 mol N =   14.01 g = 14.01 g
       mass of 4 mol H = 4(1.008 g) = 4.032 g

Chapter 8   Chemical Composition   73

		mass of 1 mol Br =	79.90 g =	79.90 g	
		molar mass of NH_4Br =		97.942 g	= 97.94 g

    d.   mass of 2 mol C =   2(12.01 g) =   24.02 g

        mass of 6 mol H =   6(1.008 g) =   6.048 g

        mass of 1 mol O =   16.00 g =   16.00 g

        molar mass of C_2H_6O =   46.07 g

    e.   mass of 2 mol H =   2(1.008 g) =   2.016 g

        mass of 1 mol S =   32.07 g =   32.07 g

        mass of 3 mol O =   3(16.00 g) =   48.00 g

        molar mass of H_2SO_3 =   82.086 g = 82.09 g

    f.   mass of 2 mol H =   2(1.008 g) =   2.016 g

        mass of 1 mol S =   32.07 g =   32.07 g

        mass of 4 mol O =   4(16.00 g) =   64.00 g

        molar mass of H_2SO_4 =   98.086 g = 98.09 g

30.   a.   mass of 2 mol N =   2(14.01 g) =   28.02 g

        mass of 8 mol H =   8(1.008 g) =   8.064 g

        mass of 1 mol S =   32.07 g =   32.07 g

        molar mass of $(NH_4)_2S$ =   68.154 g = 68.15 g

    b.   mass of 6 mol C =   6(12.01 g) =   72.06 g

        mass of 4 mol H =   4(1.008 g) =   4.032 g

        mass of 1 mol O =   16.00 g =   16.00 g

        mass of 2 mol Cl =   2(35.45 g) =   70.90 g

        molar mass of $C_6H_4OCl_2$ =   162.992 g = 162.99 g

    c.   mass of 1 mol Ba =   137.33 g =   137.33 g

        mass of 2 mol H =   2(1.008 g) =   2.016 g

        molar mass of BaH_2 =   139.346 g = 139.35 g

74   Chapter 8   Chemical Composition

d. mass of 1 mol K =   39.10 g =   39.10 g
    mass of 2 mol H =   2(1.008 g) =   2.016 g
    mass of 1 mol P =   30.97 g =   30.97 g
    mass of 4 mol O =   4(16.00 g) =   64.00 g
    ───
    molar mass of KH_2PO_4 =   136.086 g = 136.09 g

e. mass of 2 mol K =   2(39.10 g) =   78.20 g
    mass of 1 mol H =   1.008 g =   1.008 g
    mass of 1 mol P =   30.97 g =   30.97 g
    mass of 4 mol O =   4(16.00 g) =   64.00 g
    ───
    molar mass of K_2HPO_4 =   174.178 g = 174.18 g

f. mass of 3 mol K =   3(39.10 g) =   117.3 g
    mass of 1 mol P =   30.97 g =   30.97 g
    mass of 4 mol O =   4(16.00 g) =   64.00 g
    ───
    molar mass of K_3PO_4 =   212.27 g = 212.3 g

32. a. molar mass of SO_3 = 80.07 g

    $$49.2 \text{ mg } SO_3 \times \frac{1 \text{ g}}{1000 \text{ mg}} \times \frac{1 \text{ mol}}{80.07 \text{ g}} = 6.14 \times 10^{-4} \text{ mol } SO_3$$

    b. molar mass of PbO_2 = 239.2 g

    $$7.44 \times 10^4 \text{ kg } PbO_2 \times \frac{1000 \text{ g}}{1 \text{ kg}} \times \frac{1 \text{ mol}}{239.2 \text{ g}} = 3.11 \times 10^5 \text{ mol } PbO_2$$

    c. molar mass of $CHCl_3$ 119.37 g

    $$59.1 \text{ g } CHCl_3 \times \frac{1 \text{ mol}}{119.37 \text{ g}} = 0.495 \text{ mol } CHCl_3$$

    d. molar mass of $C_2H_3Cl_3$ = 133.39 g

    $$3.27 \text{ μg} \times \frac{1 \text{ g}}{10^6 \text{ μg}} \times \frac{1 \text{ mol}}{133.39 \text{ g}} = 2.45 \times 10^{-8} \text{ mol } C_2H_3Cl_3$$

e. molar mass of LiOH = 23.95 g

$$4.01 \text{ g LiOH} \times \frac{1 \text{ mol}}{23.95 \text{ g}} = 0.167 \text{ mol LiOH}$$

34. a. molar mass of NaH_2PO_4 = 120.0 g

$$4.26 \times 10^{-3} \text{ g } NaH_2PO_4 \times \frac{1 \text{ mol}}{120.0 \text{ g}} = 3.55 \times 10^{-5} \text{ mol } NaH_2PO_4$$

b. molar mass of CuCl = 99.00 g

$$521 \text{ g CuCl} \times \frac{1 \text{ mol}}{99.00 \text{ g}} = 5.26 \text{ mol CuCl}$$

c. molar mass of Fe = 55.85 g

$$151 \text{ kg Fe} \times \frac{1000 \text{ g}}{1 \text{ kg}} \times \frac{1 \text{ mol}}{55.85 \text{ g}} = 2.70 \times 10^3 \text{ mol Fe}$$

d. molar mass of SrF_2 = 125.6 g

$$8.76 \text{ g } SrF_2 \times \frac{1 \text{ mol}}{125.6 \text{ g}} = 0.0697 \text{ mol } SrF_2$$

e. molar mass of Al = 26.98 g

$$1.26 \times 10^4 \text{ g Al} \times \frac{1 \text{ mol}}{26.98 \text{ g}} = 467 \text{ mol Al}$$

36. a. molar mass of AlI_3 = 407.7 g

$$1.50 \text{ mol } AlI_3 \times \frac{407.7 \text{ g}}{1 \text{ mol}} = 612 \text{ g } AlI_3$$

b. molar mass of C_6H_6 = 78.11 g

$$1.91 \times 10^{-3} \text{ mol } C_6H_6 \times \frac{78.11 \text{ g}}{1 \text{ mol}} = 0.149 \text{ g } C_6H_6$$

c. molar mass of $C_6H_{12}O_6$ = 180.2 g

$$4.00 \text{ mol } C_6H_{12}O_6 \times \frac{180.2 \text{ g}}{1 \text{ mol}} = 721 \text{ g } C_6H_{12}O_6$$

d. molar mass of C_2H_5OH = 46.07 g

$$4.56 \times 10^5 \text{ mol } C_2H_5OH \times \frac{46.07 \text{ g}}{1 \text{ mol}} = 2.10 \times 10^7 \text{ g } C_2H_5OH$$

76   Chapter 8   *Chemical Composition*

    e.   molar mass of $Ca(NO_3)_2$ = 164.1 g

$$2.27 \text{ mol } Ca(NO_3)_2 \times \frac{164.1 \text{ g}}{1 \text{ mol}} = 373 \text{ g } Ca(NO_3)_2$$

38.   a.   molar mass of CO_2 = 44.01 g

$$1.27 \text{ mmol} \times \frac{1 \text{ mol}}{10^3 \text{ mmol}} \times \frac{44.01 \text{ g}}{1 \text{ mol}} = 0.0559 \text{ g } CO_2$$

    b.   molar mass of NCl_3 = 120.4 g

$$4.12 \times 10^3 \text{ mol } NCl_3 \times \frac{120.4 \text{ g}}{1 \text{ mol}} = 4.96 \times 10^5 \text{ g } NCl_3$$

    c.   molar mass of NH_4NO_3 = 80.05 g

$$0.00451 \text{ mol } NH_4NO_3 \times \frac{80.05 \text{ g}}{1 \text{ mol}} = 0.361 \text{ g } NH_4NO_3$$

    d.   molar mass of H_2O = 18.02 g

$$18.0 \text{ mol } H_2O \times \frac{18.02 \text{ g}}{1 \text{ mol}} = 324 \text{ g } H_2O$$

    e.   molar mass of $CuSO_4$ = 159.6 g

$$62.7 \text{ mol } CuSO_4 \times \frac{159.6 \text{ g}}{1 \text{ mol}} = 1.00 \times 10^4 \text{ g } CuSO_4$$

40.   a. $6.37 \text{ mol } CO \times \dfrac{6.022 \times 10^{23} \text{ molecules}}{1 \text{ mol}} = 3.84 \times 10^{24}$ molecules CO

    b.   molar mass of CO = 28.01 g

$$6.37 \text{ g} \times \frac{1 \text{ mol}}{28.01 \text{ g}} \times \frac{6.022 \times 10^{23} \text{ molec.}}{1 \text{ mol}} = 1.37 \times 10^{23} \text{ molecules CO}$$

    c.   molar mass of H_2O = 18.02 g

$$2.62 \times 10^{-6} \text{ g} \times \frac{6.022 \times 10^{23} \text{ molecules}}{18.02 \text{ g}} = 8.76 \times 10^{16} \text{ molecules } H_2O$$

    d. $2.62 \times 10^{-6} \text{ g} \times \dfrac{6.022 \times 10^{23} \text{ molecules}}{1 \text{ mol}} = 1.58 \times 10^{18} \text{ molecules } H_2O$

    e.   molar mass of C_6H_6 = 78.11 g

$$5.23 \text{ g} \times \frac{6.022 \times 10^{23} \text{ molecules}}{78.11 \text{ g}} = 4.03 \times 10^{22} \text{ molecules } C_6H_6$$

42. a. molar mass of Na_2SO_4 = 142.1 g

    $$2.01 \text{ g } Na_2SO_4 \times \frac{1 \text{ mol } Na_2SO_4}{142.1 \text{ g}} \times \frac{1 \text{ mol S}}{1 \text{ mol } Na_2SO_4} = 0.0141 \text{ mol S}$$

    b. molar mass of Na_2SO_3 = 126.1 g

    $$2.01 \text{ g } Na_2SO_3 \times \frac{1 \text{ mol } Na_2SO_3}{126.1 \text{ g}} \times \frac{1 \text{ mol S}}{1 \text{ mol } Na_2SO_3} = 0.0159 \text{ mol S}$$

    c. molar mass of Na_2S = 78.05 g

    $$2.01 \text{ g } Na_2S \times \frac{1 \text{ mol } Na_2S}{78.05 \text{ g}} \times \frac{1 \text{ mol S}}{1 \text{ mol } Na_2S} = 0.0258 \text{ mol S}$$

    d. molar mass of $Na_2S_2O_3$ = 158.1 g

    $$2.01 \text{ g } Na_2S_2O_3 \times \frac{1 \text{ mol } Na_2S_2O_3}{158.1 \text{ g}} \times \frac{2 \text{ mol S}}{1 \text{ mol } Na_2S_2O_3} = 0.0254 \text{ mol S}$$

44. less than

46. a. mass of Na present = 2(22.99 g) = 45.98 g
       mass of S present =   32.07 g = 32.07 g
       mass of O present = 4(16.00 g) = 64.00 g
       molar mass of Na_2SO_4 =   142.05 g

    $$\%Na = \frac{45.98 \text{ g Na}}{142.05 \text{ g}} \times 100 = 32.37\% \text{ Na}$$

    $$\%S = \frac{32.07 \text{ g S}}{142.05 \text{ g}} \times 100 = 22.58\% \text{ S}$$

    $$\%O = \frac{64.00 \text{ g O}}{142.05 \text{ g}} \times 100 = 45.05\% \text{ O}$$

    b. mass of Na present = 2(22.99 g) = 45.98 g
       mass of S present =   32.07 g = 32.07 g
       mass of O present = 3(16.00 g) = 48.00 g
       molar mass of Na_2SO_3 =   126.05 g

    $$\%Na = \frac{45.98 \text{ g Na}}{126.05 \text{ g}} \times 100 = 36.48\% \text{ Na}$$

    $$\%S = \frac{32.07 \text{ g S}}{126.05 \text{ g}} \times 100 = 25.44\% \text{ S}$$

    $$\%O = \frac{48.00 \text{ g O}}{126.05 \text{ g}} \times 100 = 38.08\% \text{ O}$$

78 Chapter 8 Chemical Composition

c. mass of Na present = $2(22.99 \text{ g})$ = 45.98 g
    mass of S present = 32.07 g = 32.07 g
    ───
    molar mass of Na_2S = 78.05 g

    $\%Na = \dfrac{45.98 \text{ g Na}}{78.05 \text{ g}} \times 100 = 58.91\% \text{ Na}$

    $\%S = \dfrac{32.07 \text{ g S}}{78.05 \text{ g}} \times 100 = 41.09\% \text{ S}$

d. mass of Na present = $2(22.99 \text{ g})$ = 45.98 g
    mass of S present = $2(32.07 \text{ g})$ = 64.14 g
    mass of O present = $3(16.00 \text{ g})$ = 48.00 g
    ───
    molar mass of $Na_2S_2O_3$ = 158.12 g

    $\%Na = \dfrac{45.98 \text{ g Na}}{158.12 \text{ g}} \times 100 = 29.08\% \text{ Na}$

    $\%S = \dfrac{64.14 \text{ g S}}{158.12 \text{ g}} \times 100 = 40.56\% \text{ S}$

    $\%O = \dfrac{48.00 \text{ g O}}{158.12 \text{ g}} \times 100 = 30.36\% \text{ O}$

e. mass of K present = $3(39.10 \text{ g})$ = 117.3 g
    mass of P present = 30.97 g = 30.97 g
    mass of O present = $4(16.00 \text{ g})$ = 64.00 g
    ───
    molar mass of K_3PO_4 = 212.3 g

    $\%K = \dfrac{117.3 \text{ g K}}{212.3 \text{ g}} \times 100 = 55.25\% \text{ K}$

    $\%P = \dfrac{30.97 \text{ g P}}{212.3 \text{ g}} \times 100 = 14.59\% \text{ P}$

    $\%O = \dfrac{64.00 \text{ g O}}{212.3 \text{ g}} \times 100 = 30.15\% \text{ O}$

f. mass of K present = $2(39.10 \text{ g})$ = 78.20 g
    mass of H present = 1.008 g = 1.008 g
    mass of P present = 30.97 g = 30.97 g
    mass of O present = $4(16.00 \text{ g})$ = 64.00 g
    ───
    molar mass of K_2HPO_4 = 174.178 g = 174.18 g

    $\%K = \dfrac{78.20 \text{ g K}}{174.18 \text{ g}} \times 100 = 44.90\% \text{ K}$

$$\%H = \frac{1.008 \text{ g H}}{174.18 \text{ g}} \times 100 = 0.5787\% \text{ H}$$

$$\%P = \frac{30.97 \text{ g P}}{174.18 \text{ g}} \times 100 = 17.78\% \text{ P}$$

$$\%O = \frac{64.00 \text{ g O}}{174.18 \text{ g}} \times 100 = 36.74\% \text{ O}$$

g.
mass of K present =	39.10 g =	39.10 g
mass of H present =	2(1.008 g) =	2.016 g
mass of P present =	30.97 g =	30.97 g
mass of O present =	4(16.00 g) =	64.00 g
molar mass of KH_2PO_4 =		136.09 g

$$\%K = \frac{39.10 \text{ g K}}{136.09 \text{ g}} \times 100 = 28.73\% \text{ K}$$

$$\%H = \frac{2.016 \text{ g H}}{136.09 \text{ g}} \times 100 = 1.481\% \text{ H}$$

$$\%P = \frac{30.97 \text{ g P}}{136.09 \text{ g}} \times 100 = 22.76\% \text{ P}$$

$$\%O = \frac{64.00 \text{ g O}}{136.09 \text{ g}} \times 100 = 47.03\% \text{ O}$$

h.
mass of K present =	3(39.10 g) =	117.3 g
mass of P present =	30.97 g =	30.97 g
molar mass of K_3P =		148.27 g = 148.3 g

$$\%K = \frac{117.3 \text{ g K}}{148.3 \text{ g}} \times 100 = 79.10\% \text{ K}$$

$$\%P = \frac{30.97 \text{ g P}}{148.3 \text{ g}} \times 100 = 20.88\% \text{ P}$$

48. a. molar mass of $CuBr_2$ = 223.4

$$\% \text{Cu} = \frac{63.55 \text{ g Cu}}{223.4 \text{ g}} \times 100 = 28.45\% \text{ Cu}$$

b. molar mass of CuBr = 143.5 g

$$\% \text{Cu} = \frac{63.55 \text{ g Cu}}{143.5 \text{ g}} \times 100 = 44.29\% \text{ Cu}$$

80   Chapter 8   Chemical Composition

c. molar mass of $FeCl_2$ = 126.75 g

$$\%\text{ Fe} = \frac{55.85 \text{ g Fe}}{126.75 \text{ g}} \times 100 = 44.06\% \text{ Fe}$$

d. molar mass of $FeCl_3$ = 162.2 g

$$\%\text{ Fe} = \frac{55.85 \text{ g Fe}}{162.2 \text{ g}} \times 100 = 34.43\% \text{ Fe}$$

e. molar mass of CoI_2 = 312.7 g

$$\%\text{ Co} = \frac{58.93 \text{ g Co}}{312.7 \text{ g}} \times 100 = 18.85\% \text{ Co}$$

f. molar mass of CoI_3 = 439.6 g

$$\%\text{ Co} = \frac{58.93 \text{ g Co}}{439.6 \text{ g}} \times 100 = 13.41\% \text{ Co}$$

g. molar mass of SnO = 134.7 g

$$\%\text{ Sn} = \frac{118.7 \text{ g Sn}}{134.7 \text{ g}} \times 100 = 88.12\% \text{ Sn}$$

h. molar mass of SnO_2 = 150.7 g

$$\%\text{ Sn} = \frac{118.7 \text{ g Sn}}{150.7 \text{ g}} \times 100 = 78.77\% \text{ Sn}$$

50. a. molar mass of $FeCl_3$ = 162.2 g

$$\%\text{Fe} = \frac{55.85 \text{ g Fe}}{162.2 \text{ g}} \times 100 = 34.43\% \text{ Fe}$$

b. molar mass of OF_2 = 54.00 g

$$\%\text{O} = \frac{16.00 \text{ g O}}{54.00 \text{ g}} \times 100 = 29.63\% \text{ O}$$

c. molar mass of C_6H_6 = 78.11 g

$$\%\text{C} = \frac{72.06 \text{ g C}}{78.11 \text{ g}} \times 100 = 92.25\% \text{ C}$$

d. molar mass of NH_4ClO_4 = 117.5 g

$$\%\text{N} = \frac{14.01 \text{ g N}}{117.5 \text{ g}} \times 100 = 11.92\% \text{ N}$$

e. molar mass of Ag_2O = 231.8 g

$$\%\text{Ag} = \frac{215.8 \text{ g Ag}}{231.8 \text{ g}} \times 100 = 93.10\% \text{ Ag}$$

f. molar mass of $CoCl_2$ = 129.83 g

$$\%Co = \frac{58.93 \text{ g Co}}{129.83 \text{ g}} \times 100 = 45.39\% \text{ Co}$$

g. molar mass of N_2O_4 = 92.02 g

$$\%N = \frac{28.02 \text{ g N}}{92.02 \text{ g}} \times 100 = 30.45\% \text{ N}$$

h. molar mass of $MnCl_2$ = 125.84 g

$$\%Mn = \frac{54.94 \text{ g Mn}}{125.8 \text{ g}} \times 100 = 43.66\% \text{ Mn}$$

52. a. molar mass of NH_4Cl = 53.49 g

molar mass of NH_4^+ ion = 18.04 g

$$\% NH_4^+ = \frac{18.04 \text{ g } NH_4^+}{53.49 \text{ g}} \times 100 = 33.73\% \text{ } NH_4^+$$

b. molar mass of $CuSO_4$ = 159.62 g

molar mass of Cu^{2+} = 63.55 g

$$\% Cu^{2+} = \frac{63.55 \text{ g } Cu^{2+}}{159.62 \text{ g}} \times 100 = 39.81\% \text{ } Cu^{2+}$$

c. molar mass of $AuCl_3$ = 303.4 g

molar mass of Au^{3+} ion = 197.0 g

$$\% Au^{3+} = \frac{197.0 \text{ g } Au^{3+}}{303.4 \text{ g}} \times 100 = 64.93\% \text{ } Au^{3+}$$

d. molar mass of $AgNO_3$ = 169.9 g

molar mass of Ag^+ ion = 107.9 g

$$\% Ag^+ = \frac{107.9 \text{ g } Ag^+}{169.9 \text{ g}} \times 100 = 63.51\% \text{ } Ag^+$$

54. The empirical formula represents the smallest whole number ratio of the elements present in a compound. The molecular formula indicates the actual number of atoms of each element found in a molecule of the substance.

56. a. yes (each of these has empirical formula CH)
    b. no (the number of hydrogen atoms is wrong)
    c. yes (both have empirical formula NO_2)
    d. no (the number of hydrogen and oxygen atoms is wrong)

82   Chapter 8   Chemical Composition

58. $2.514 \text{ g Ca} \times \dfrac{1 \text{ mol}}{40.08 \text{ g Ca}} = 0.06272 \text{ mol Ca}$

The increase in mass represents the oxygen with which the calcium reacted:

$1.004 \text{ g O} \times \dfrac{1 \text{ mol}}{16.00 \text{ g O}} = 0.06275 \text{ mol O}$

Since we have effectively the same number of moles of Ca and O, the empirical formula must be CaO.

60.   Consider having 100.0 g of the compound. Then the percentages of the elements present are numerically equal to their masses in grams.

$58.84 \text{ g Ba} \times \dfrac{1 \text{ mol}}{137.3 \text{ g Ba}} = 0.4286 \text{ mol Ba}$

$13.74 \text{ g S} \times \dfrac{1 \text{ mol}}{32.07 \text{ g S}} = 0.4284 \text{ mol S}$

$27.43 \text{ g O} \times \dfrac{1 \text{ mol}}{16.00 \text{ g O+}} = 1.714 \text{ mol O}$

Dividing each number of moles by the smallest number of moles (0.4284 mol S) gives

$\dfrac{0.4286 \text{ mol Ba}}{0.4284} = 1.000 \text{ mol Ba}$

$\dfrac{0.4284 \text{ mol S}}{0.4284} = 1.000 \text{ mol S}$

$\dfrac{1.714 \text{ mol O}}{0.4284} = 4.001 \text{ mol O}$

The empirical formula is $BaSO_4$.

62.   Consider 100.0 g of the compound.

$29.16 \text{ g N} \times \dfrac{1 \text{ mol}}{14.01 \text{ g N}} = 2.081 \text{ mol N}$

$8.392 \text{ g H} \times \dfrac{1 \text{ mol}}{1.008 \text{ g H}} = 8.325 \text{ mol H}$

$12.50 \text{ g C} \times \dfrac{1 \text{ mol}}{12.01 \text{ g C}} = 1.041 \text{ mol C}$

$49.95 \text{ g O} \times \dfrac{1 \text{ mol}}{16.00 \text{ g O}} = 3.122 \text{ mol O}$

Dividing each number of moles by the smallest (1.041 mol C) gives

$$\frac{2.081 \text{ mol N}}{1.041} = 1.999 \text{ mol N}$$

$$\frac{8.325 \text{ mol H}}{1.041} = 7.997 \text{ mol H}$$

$$\frac{1.041 \text{ mol C}}{1.041} = 1.000 \text{ mol C}$$

$$\frac{3.121 \text{ mol O}}{1.041} = 2.998 \text{ mol O}$$

The empirical formula is $N_2H_8CO_3$ [i.e, $(NH_4)_2CO_3$].

64. Consider 100.0 g of the compound.

$$55.06 \text{ g Co} \times \frac{1 \text{ mol}}{58.93 \text{ g Co}} = 0.9343 \text{ mol Co}$$

If the sulfide of cobalt is 55.06% Co, then it is 44.94% S by mass.

$$44.94 \text{ g S} \times \frac{1 \text{ mol}}{32.07 \text{ g S}} = 1.401 \text{ mol S}$$

Dividing each number of moles by the smaller (0.9343 mol Co) gives

$$\frac{0.9343 \text{ mol Co}}{0.9343} = 1.000 \text{ mol Co}$$

$$\frac{1.401 \text{ mol S}}{0.9343} = 1.500 \text{ mol S}$$

Multiplying by two, to convert to whole numbers of moles, gives the empirical formula for the compound as Co_2S_3.

66. $$10.00 \text{ g Cu} \times \frac{1 \text{ mol}}{63.55 \text{ g Cu}} = 0.1574 \text{ mol Cu}$$

$$2.52 \text{ g O} \times \frac{1 \text{ mol}}{16.00 \text{ g O}} = 0.158 \text{ mol O}$$

The numbers of moles are almost equal: the empirical formula is just CuO.

68. Consider 100.0 g of the compound.

$$32.13 \text{ g Al} \times \frac{1 \text{ mol}}{26.98 \text{ g Al}} = 1.191 \text{ mol Al}$$

84   Chapter 8   Chemical Composition

$$67.87 \text{ g F} \times \frac{1 \text{ mol}}{19.00 \text{ g F}} = 3.572 \text{ mol F}$$

Dividing each number of moles by the smaller number (1.191 mol Al) gives

$$\frac{1.191 \text{ mol Al}}{1.191} = 1.000 \text{ mol Al}$$

$$\frac{3.572 \text{ mol F}}{1.191} = 2.999 \text{ mol F}$$

The empirical formula is AlF_3

70. Consider 100.0 g of the compound.

$$59.78 \text{ g Li} \times \frac{1 \text{ mol Li}}{6.941 \text{ g Li}} = 8.613 \text{ mol Li}$$

$$40.22 \text{ g N} \times \frac{1 \text{ mol N}}{14.01 \text{ g N}} = 2.871 \text{ mol N}$$

Dividing each number of moles by the smaller number of moles (2.871 mol N) gives

$$\frac{8.613 \text{ mol Li}}{2.871} = 3.000 \text{ mol Li}$$

$$\frac{2.871 \text{ mol N}}{2.871} = 1.000 \text{ mol N}$$

The empirical formula is Li_3N.

72. Consider 100.0 g of the compound.

$$15.77 \text{ g Al} \times \frac{1 \text{ mol Al}}{26.98 \text{ g Al}} = 0.5845 \text{ mol Al}$$

$$28.11 \text{ g S} \times \frac{1 \text{ mol S}}{32.07 \text{ g S}} = 0.8765 \text{ mol S}$$

$$56.12 \text{ g O} \times \frac{1 \text{ mol O}}{16.00 \text{ g O}} = 3.508 \text{ mol O}$$

Dividing each number of moles by the smallest number of moles (0.5845 mol Al) gives

$$\frac{0.5845 \text{ mol Al}}{0.5845} = 1.000 \text{ mol Al}$$

$$\frac{0.8765 \text{ mol S}}{0.5845} = 1.500 \text{ mol S}$$

$$\frac{3.508 \text{ mol O}}{0.5845} = 6.002 \text{ mol O}$$

Multiplying these relative numbers of moles by 2 to give whole numbers gives the empirical formula as $Al_2S_3O_{12}$ [i.e., $Al_2(SO_4)_3$].

74. Compound 1: Assume 100.0 g of the compound.

    $$83.12 \text{ g Na} \times \frac{1 \text{ mol Na}}{22.99 \text{ g Na}} = 3.615 \text{ mol Na}$$

    $$16.88 \text{ g N} \times \frac{1 \text{ mol N}}{14.01 \text{ g N}} = 1.205 \text{ mol Na}$$

    Dividing each number of moles by the smaller (1.205 mol Na) indicates that the formula of Compound 1 is Na_3N.

    Compound 2: Assume 100.0 g of the compound.

    $$35.36 \text{ g Na} \times \frac{1 \text{ mol Na}}{22.99 \text{ g Na}} = 1.538 \text{ mol Na}$$

    $$64.64 \text{ g N} \times \frac{1 \text{ mol N}}{14.01 \text{ g N}} = 4.614 \text{ mol N}$$

    Dividing each number of moles by the smaller (1.538 mol Na) indicates that the formula of Compound 2 is NaN_3

76. If only the empirical formula is known, the molar mass of the substance must be determined before the molecular formula can be calculated.

78. empirical formula mass of CH = 13 g

    $$n = \frac{\text{molar mass}}{\text{empirical formula mass}} = \frac{78 \text{ g}}{13 \text{ g}} = 6$$

    The molecular formula is $(CH)_6$ or C_6H_6.

80. empirical formula mass of CH_4O = 32.04 g

    $$n = \frac{\text{molar mass}}{\text{empirical formula mass}} = \frac{192 \text{ g}}{32.04 \text{ g}} = 6$$

    molecular formula is $(CH_4O)_6 = C_6H_{24}O_6$

82. Consider 100.0 g of the compound.

    $$65.45 \text{ g C} \times \frac{1 \text{ mol C}}{12.01 \text{ g C}} = 5.450 \text{ mol C}$$

    $$5.492 \text{ g H} \times \frac{1 \text{ mol H}}{1.008 \text{ g H}} = 5.448 \text{ mol H}$$

86   Chapter 8   Chemical Composition

$$29.06 \text{ g O} \times \frac{1 \text{ mol O}}{16.00 \text{ g O}} = 1.816 \text{ mol O}$$

Dividing each number of moles by the smallest number of moles (1.816 mol O) gives

$$\frac{5.450 \text{ mol C}}{1.816} = 3.001 \text{ mol C}$$

$$\frac{5.448 \text{ mol H}}{1.816} = 3.000 \text{ mol H}$$

$$\frac{1.816 \text{ mol O}}{1.816} = 1.000 \text{ mol O}$$

The empirical formula is C_3H_3O, and the empirical formula mass is approximately 55 g.

$$n = \frac{\text{molar mass}}{\text{empirical formula mass}} = \frac{110 \text{ g}}{55 \text{ g}} = 2$$

The molecular formula is $(C_3H_3O)_2 = C_6H_6O_2$.

84.
5.00 g Al	0.185 mol	1.12×10^{23} atoms
0.140 g Fe	0.00250 mol	1.51×10^{21} atoms
2.7×10^2 g Cu	4.3 mol	2.6×10^{24} atoms
0.00250 g Mg	1.03×10^{-4} mol	6.19×10^{19} atoms
0.062 g Na	2.7×10^{-3} mol	1.6×10^{21} atoms
3.95×10^{-18} g U	1.66×10^{-20} mol	1.00×10^4 atoms

86. mass of 2 mol X =   2(41.2 g) =   82.4 g

mass of 1 mol Y =   57.7 g =   57.7 g

mass of 3 mol Z =   3(63.9 g) =   191.7 g

---

molar mass of X_2YZ_3 =   331.8 g

$$\% \text{ X} = \frac{82.4 \text{ g}}{331.8 \text{ g}} \times 100 = 24.8\% \text{ X}$$

$$\% \text{ Y} = \frac{57.7 \text{ g}}{331.8 \text{ g}} \times 100 = 17.4\% \text{ Y}$$

$$\% \text{ Z} = \frac{191.7 \text{ g}}{331.8 \text{ g}} \times 100 = 57.8\% \text{ Z}$$

If the molecular formula were actually $X_4Y_2Z_6$, the percentage composition would be the same, and the *relative* mass of each element present would not change. The molecular formula is always a whole number multiple of the empirical formula.

88. For the first compound (*restricted* amount of oxygen)

$$2.118 \text{ g Cu} \times \frac{1 \text{ mol Cu}}{63.54 \text{ g Cu}} = 0.03333 \text{ mol Cu}$$

$$0.2666 \text{ g O} \times \frac{1 \text{ mol O}}{16.00 \text{ g O}} = 0.01666 \text{ mol O}$$

Since the number of moles of Cu (0.03333 mol) is twice the number of moles of O (0.01666 mol), the empirical formula is Cu_2O.

For the second compound (stream of pure oxygen)

$$2.118 \text{ g Cu} \times \frac{1 \text{ mol Cu}}{63.54 \text{ g Cu}} = 0.03333 \text{ mol Cu}$$

$$0.5332 \text{ g O} \times \frac{1 \text{ mol O}}{16.00 \text{ g O}} = 0.03333 \text{ mol O}$$

Since the numbers of moles are the same, the empirical formula is CuO.

90. a. molar mass H_2O = 18.02 g

$$4.21 \text{ g} \times \frac{1 \text{ mol}}{18.02 \text{ g}} \times \frac{6.022 \times 10^{23} \text{ molecules}}{1 \text{ mol}} = 1.41 \times 10^{23} \text{ molecules}$$

The sample contains 1.41×10^{23} oxygen atoms and $2(1.41 \times 10^{23})$ = 2.82×10^{23} hydrogen atoms

b. molar mass CO_2 = 44.01 g

$$6.81 \text{ g} \times \frac{1 \text{ mol}}{44.01 \text{ g}} \times \frac{6.022 \times 10^{23} \text{ molecules}}{1 \text{ mol}} = 9.32 \times 10^{22} \text{ molecules}$$

The sample contains 9.32×10^{22} carbon atoms and $2(9.32 \times 10^{22})$ = 1.86×10^{23} oxygen atoms.

c. molar mass C_6H_6 = 78.11 g

$$0.000221 \text{ g} \times \frac{1 \text{ mol}}{78.11 \text{ g}} \times \frac{6.022 \times 10^{23} \text{ molecules}}{1 \text{ mol}} = 1.70 \times 10^{18} \text{ molec.}$$

The sample contains $6(1.70 \times 10^{18}) = 1.02 \times 10^{19}$ atoms of each element.

d. $2.26 \text{ mol} \times \dfrac{6.022 \times 10^{23} \text{ molecules}}{1 \text{ mol}} = 1.36 \times 10^{24}$ molecules

atoms C = $12(1.36 \times 10^{24}) = 1.63 \times 10^{25}$ atoms
atoms H = $22(1.36 \times 10^{24}) = 2.99 \times 10^{25}$ atoms
atoms O = $11(1.36 \times 10^{24}) = 1.50 \times 10^{25}$ atoms

92. a. molar mass of C_3O_2 = 3(12.01 g) + 2(16.00 g) = 68.03 g

$$\% \text{ C} = \frac{36.03 \text{ g}}{68.03 \text{ g}} = 52.96\% \text{ C}$$

$$7.819 \text{ g } C_3O_2 \times \frac{52.96 \text{ g C}}{100.0 \text{ g } C_3O_2} = 4.141 \text{ g C}$$

$$4.141 \text{ g C} \times \frac{6.022 \times 10^{23} \text{ C atoms}}{12.01 \text{ g C}} = 2.076 \times 10^{23} \text{ C atoms}$$

b. molar mass of CO = 12.01 g + 16.00 g = 28.01 g

$$\% \text{ C} = \frac{12.01 \text{ g}}{28.01 \text{ g}} \times 100 = 42.88\% \text{ C}$$

$$1.53 \times 10^{21} \text{ molecules CO} \times \frac{1 \text{ C atom}}{1 \text{ molecule CO}} = 1.53 \times 10^{21} \text{ C atoms}$$

$$1.53 \times 10^{21} \text{ C atoms} \times \frac{12.01 \text{ g C}}{6.022 \times 10^{23} \text{ C atoms}} = 0.0305 \text{ g C}$$

c. molar mass of C_6H_6O = 6(12.01 g) + 6(1.008 g) + 16.00 g = 94.11 g

$$\% \text{ C} = \frac{72.06 \text{ g}}{94.11 \text{ g}} \times 100 = 76.57\% \text{ C}$$

$$0.200 \text{ mol } C_6H_6O \times \frac{6 \text{ mol C}}{1 \text{ mol } C_6H_6O} = 1.20 \text{ mol C}$$

$$1.20 \text{ mol C} \times \frac{12.01 \text{ g C}}{1 \text{ mol C}} = 14.4 \text{ g C}$$

$$14.4 \text{ g C} \times \frac{6.022 \times 10^{23} \text{ C atoms}}{12.01 \text{ g C}} = 7.22 \times 10^{23} \text{ C atoms}$$

94. $2.24 \text{ g Co} \times \dfrac{55.85 \text{ g Fe}}{58.93 \text{ g Co}} = 2.12 \text{ g Fe}$

96. $5.00 \text{ g Te} \times \dfrac{200.6 \text{ g Hg}}{127.6 \text{ g Te}} = 7.86 \text{ g Hg}$

98. $153.8 \text{ g } CCl_4 = 6.022 \times 10^{23}$ molecules CCl_4

   $1 \text{ molecule} \times \dfrac{153.8 \text{ g}}{6.022 \times 10^{23} \text{ molecules}} = 2.554 \times 10^{-22} \text{ g}$

100. a. molar mass of $C_2H_5O_2N = 75.07$ g

   mass fraction N $= \dfrac{14.01 \text{ g}}{75.07 \text{ g}}$

   $5.000 \text{ g} \times \dfrac{14.01 \text{ g}}{75.07 \text{ g}} = 0.9331 \text{ g N}$

   b. molar mass of $Mg_3N_2 = 3(24.31 \text{ g}) + 2(14.01 \text{ g}) = 100.95$ g

   mass fraction N $= \dfrac{28.02 \text{ g}}{100.95 \text{ g}}$

   $5.000 \text{ g} \times \dfrac{28.02 \text{ g}}{100.95 \text{ g}} = 1.388 \text{ g N}$

   c. molar mass of $Ca(NO_3)_2 =$
   $40.08 \text{ g} + 2(14.01 \text{ g}) + 6(16.00 \text{ g}) = 164.10$ g

   mass fraction N $= \dfrac{28.02 \text{ g}}{164.10 \text{ g}}$

   $5.000 \text{ g} \times \dfrac{28.02 \text{ g}}{164.10 \text{ g}} = 0.8537 \text{ g N}$

   d. molar mass of $N_2O_4 = 2(14.01 \text{ g}) + 4(16.00 \text{ g}) = 92.02$ g

   mass fraction N $= \dfrac{28.02 \text{ g}}{92.02 \text{ g}}$

   $5.000 \text{ g} \times \dfrac{28.02 \text{ g}}{92.02 \text{ g}} = 1.522 \text{ g N}$

102. Consider 100.0 g of the compound.

   $16.39 \text{ g Mg} \times \dfrac{1 \text{ mol Mg}}{24.31 \text{ g Mg}} = 0.6742 \text{ mol Mg}$

$$18.89 \text{ g N} \times \frac{1 \text{ mol N}}{14.01 \text{ g N}} = 1.348 \text{ mol N}$$

$$64.72 \text{ g O} \times \frac{1 \text{ mol O}}{16.00 \text{ g O}} = 4.045 \text{ mol O}$$

Dividing each number of moles by the smallest number of moles

$$\frac{0.6742 \text{ mol Mg}}{0.6742} = 1.000 \text{ mol Mg}$$

$$\frac{1.348 \text{ mol N}}{0.6742} = 1.999 \text{ mol N}$$

$$\frac{4.045 \text{ mol O}}{0.6742} = 5.999 \text{ mol O}$$

The empirical formula is MgN_2O_6 [i.e., $Mg(NO_3)_2$].

104. We use the *average* mass because this average is a *weighted average* and takes into account both the masses and the relative abundances of the various isotopes.

106. $1.98 \times 10^{13} \text{ amu} \times \frac{1 \text{ Na atom}}{22.99 \text{ amu}} = 8.61 \times 10^{11} \text{ Na atoms}$

$3.01 \times 10^{23} \text{ Na atoms} \times \frac{22.99 \text{ amu}}{1 \text{ Na atom}} = 6.92 \times 10^{24} \text{ amu}$

108. a. $5.0 \text{ mol K} \times \frac{39.10 \text{ g}}{1 \text{ mol}} = 195 \text{ g} = 2.0 \times 10^2 \text{ g K}$

b. $0.000305 \text{ mol Hg} \times \frac{200.6 \text{ g}}{1 \text{ mol}} = 0.0612 \text{ g Hg}$

c. $2.31 \times 10^{-5} \text{ mol Mn} \times \frac{54.94 \text{ g}}{1 \text{ mol}} = 1.27 \times 10^{-3} \text{ g Mn}$

d. $10.5 \text{ mol P} \times \frac{30.97 \text{ g}}{1 \text{ mol}} = 325 \text{ g P}$

e. $4.9 \times 10^4 \text{ mol Fe} \times \frac{55.85 \text{ g}}{1 \text{ mol}} = 2.7 \times 10^6 \text{ g Fe}$

f. $125 \text{ mol Li} \times \frac{6.941 \text{ g}}{1 \text{ mol}} = 868 \text{ g Li}$

g. $0.01205 \text{ mol F} \times \frac{19.00 \text{ g}}{1 \text{ mol}} = 0.2290 \text{ g F}$

Chapter 8   Chemical Composition   91

110. a. mass of 1 mol Fe = 1(55.85 g) = 55.85 g
         mass of 1 mol S = 1(32.07 g) = 32.07 g
         mass of 4 mol O = 4(16.00 g) = 64.00 g
         ───────────────────────────────────────
         molar mass of $FeSO_4$ =   151.92 g

   b. mass of 1 mol Hg = 1(200.6 g) = 200.6 g
       mass of 2 mol I = 2(126.9 g) = 253.8 g
       ───────────────────────────────────────
       molar mass of HgI_2 =   454.4 g

   c. mass of 1 mol Sn = 1(118.7 g) = 118.7 g
       mass of 2 mol O = 2(16.00 g) = 32.00 g
       ───────────────────────────────────────
       molar mass of SnO_2 =   150.7 g

   d. mass of 1 mol Co = 1(58.93 g) = 58.93 g
       mass of 2 mol Cl = 2(35.45 g) = 70.90 g
       ───────────────────────────────────────
       molar mass of $CoCl_2$ =   129.83 g

   e. mass of 1 mol Cu = 1(63.55 g) = 63.55 g
       mass of 2 mol N = 2(14.01 g) = 28.02 g
       mass of 6 mol O = 6(16.00 g) = 96.00 g
       ───────────────────────────────────────
       molar mass of $Cu(NO_3)_2$ =   187.57 g

112. a. molar mass of $(NH_4)_2S$ = 68.15 g

         $$21.2 \text{ g} \times \frac{1 \text{ mol}}{68.15 \text{ g}} = 0.311 \text{ mol } (NH_4)_2S$$

   b. molar mass of $Ca(NO_3)_2$ = 164.1 g

         $$44.3 \text{ g} \times \frac{1 \text{ mol}}{164.1 \text{ g}} = 0.270 \text{ mol } Ca(NO_3)_2$$

   c. molar mass of Cl_2O = 86.9 g

         $$4.35 \text{ g} \times \frac{1 \text{ mol}}{86.9 \text{ g}} = 0.0501 \text{ mol } Cl_2O$$

   d. 1.0 lb = 454 g
       molar mass of $FeCl_3$ = 162.2

         $$454 \text{ g} \times \frac{1 \text{ mol}}{162.2 \text{ g}} = 2.8 \text{ mol } FeCl_3$$

92   Chapter 8   Chemical Composition

e. $1.0 \text{ kg} = 1.0 \times 10^3 \text{ g}$

molar mass of $FeCl_3 = 162.2$ g

$1.0 \times 10^3 \text{ g} \times \dfrac{1 \text{ mol}}{162.2 \text{ g}} = 6.2 \text{ mol } FeCl_3$

114. a.   molar mass of $CuSO_4 = 159.62$ g

$2.6 \times 10^{-2} \text{ mol} \times \dfrac{159.62 \text{ g}}{1 \text{ mol}} = 4.2 \text{ g } CuSO_4$

b.   molar mass of $C_2F_4 = 100.0$ g

$3.05 \times 10^3 \text{ mol} \times \dfrac{100.0 \text{ g}}{1 \text{ mol}} = 3.05 \times 10^5 \text{ g } C_2F_4$

c.   7.83 mmol = 0.00783 mol

molar mass of $C_5H_8 = 68.11$ g

$0.00783 \text{ mol} \times \dfrac{68.11 \text{ g}}{1 \text{ mol}} = 0.533 \text{ g } C_5H_8$

d.   molar mass of $BiCl_3 = 315.3$ g

$6.30 \text{ mol} \times \dfrac{315.3 \text{ g}}{1 \text{ mol}} = 1.99 \times 10^3 \text{ g } BiCl_3$

e.   molar mass of $C_{12}H_{22}O_{11} = 342.3$ g

$12.2 \text{ mol} \times \dfrac{342.3 \text{ g}}{1 \text{ mol}} = 4.18 \times 10^3 \text{ g } C_{12}H_{22}O_{11}$

116. a.   molar mass of $C_6H_{12}O_6 = 180.2$ g

$3.45 \text{ g} \times \dfrac{6.022 \times 10^{23} \text{ molecules}}{180.2 \text{ g}} = 1.15 \times 10^{22} \text{ molecules } C_6H_{12}O_6$

b. $3.45 \text{ mol} \times \dfrac{6.022 \times 10^{23} \text{ molecules}}{1 \text{ mol}} = 2.08 \times 10^{24} \text{ molecules } C_6H_{12}O_6$

c.   molar mass of $ICl_5 = 304.2$ g

$25.0 \text{ g} \times \dfrac{6.022 \times 10^{23} \text{ molecules}}{304.2 \text{ g}} = 4.95 \times 10^{22} \text{ molecules } ICl_5$

d.   molar mass of $B_2H_6 = 27.67$ g

$1.00 \text{ g} \times \dfrac{6.022 \times 10^{23} \text{ molecules}}{27.67 \text{ g}} = 2.18 \times 10^{22} \text{ molecules } B_2H_6$

e.   1.05 mmol = 0.00105 mol

$$0.00105 \text{ mol} \times \frac{6.022 \times 10^{23} \text{ form. units}}{1 \text{ mol}} = 6.32 \times 10^{20} \text{ form. units}$$

118. a. mass of Ca present =   3(40.08 g) =   120.24 g
       mass of P present =   2(30.97 g) =   61.94 g
       mass of O present =   8(16.00 g) =   128.00 g
       ──
       molar mass of $Ca_3(PO_4)_2$ =   310.18 g

    $\% \text{ Ca} = \dfrac{120.24 \text{ g Ca}}{310.18 \text{ g}} \times 100 = 38.76\% \text{ Ca}$

    $\% \text{ P} = \dfrac{61.94 \text{ g P}}{310.18 \text{ g}} \times 100 = 19.97\% \text{ P}$

    $\% \text{ O} = \dfrac{128.00 \text{ g O}}{310.18 \text{ g}} \times 100 = 41.27\% \text{ O}$

   b. mass of Cd present =   112.4 g =   112.4 g
       mass of S present =   32.07 g =   32.07 g
       mass of O present =   4(16.00 g) =   64.00 g
       ──
       molar mass of $CdSO_4$ =   208.5 g

    $\% \text{ Cd} = \dfrac{112.4 \text{ g Cd}}{208.5 \text{ g}} \times 100 = 53.91\% \text{ Cd}$

    $\% \text{ S} = \dfrac{32.07 \text{ g S}}{208.5 \text{ g}} \times 100 = 15.38\% \text{ S}$

    $\% \text{ O} = \dfrac{64.00 \text{ g O}}{208.5 \text{ g}} \times 100 = 30.70\% \text{ O}$

   c. mass of Fe present =   2(55.85 g) =   111.7 g
       mass of S present =   3(32.07 g) =   96.21 g
       mass of O present =   12(16.00 g) =   192.0 g
       ──
       molar mass of $Fe_2(SO_4)_3$ =   399.9 g

    $\% \text{ Fe} = \dfrac{111.7 \text{ g Fe}}{399.9 \text{ g}} \times 100 = 27.93\% \text{ Fe}$

    $\% \text{ S} = \dfrac{96.21 \text{ g S}}{399.9 \text{ g}} \times 100 = 24.06\% \text{ S}$

    $\% \text{ O} = \dfrac{192.0 \text{ g O}}{399.9 \text{ g}} \times 100 = 48.01\% \text{ O}$

94   Chapter 8   Chemical Composition

d.   mass of Mn present =   54.94 g =   54.94 g
     mass of Cl present =   2(35.45 g) =   70.90 g
     ───
     molar mass of $MnCl_2$ =   125.84 g

   % Mn = $\dfrac{54.94 \text{ g Mn}}{125.84 \text{ g}}$ × 100 = 43.66% Mn

   % Cl = $\dfrac{70.90 \text{ g Cl}}{125.84 \text{ g}}$ × 100 = 56.34% Cl

e.   mass of N present =   2(14.01 g) =   28.02 g
     mass of H present =   8(1.008 g) =   8.064 g
     mass of C present =   12.01 g =   12.01 g
     mass of O present =   3(16.00 g) =   48.00 g
     ───
     molar mass of $(NH_4)_2CO_3$ =   96.09 g

   % N = $\dfrac{28.02 \text{ g N}}{96.09 \text{ g}}$ × 100 = 29.16% N

   % H = $\dfrac{8.064 \text{ g H}}{96.09 \text{ g}}$ × 100 = 8.392% H

   % C = $\dfrac{12.01 \text{ g C}}{96.09 \text{ g}}$ × 100 = 12.50% C

   % O = $\dfrac{48.00 \text{ g O}}{96.09 \text{ g}}$ × 100 = 49.95% O

f.   mass of Na present =   22.99 g =   22.99 g
     mass of H present =   1.008 g =   1.008 g
     mass of C present =   12.01 g =   12.01 g
     mass of O present =   3(16.00 g) =   48.00 g
     ───
     molar mass of $NaHCO_3$ =   84.01 g

   % Na = $\dfrac{22.99 \text{ g Na}}{84.01 \text{ g}}$ × 100 = 27.37% Na

   % H = $\dfrac{1.008 \text{ g H}}{84.01 \text{ g}}$ × 100 = 1.200% H

   % C = $\dfrac{12.01 \text{ g C}}{84.01 \text{ g}}$ × 100 = 14.30% C

   % O = $\dfrac{48.00 \text{ g O}}{84.01 \text{ g}}$ × 100 = 57.14% O

g. mass of C present =   12.01 g =   12.01 g
    mass of O present =   2(16.00 g) =   32.00 g
    ──
    molar mass of CO_2 =   44.01 g

    % C = $\dfrac{12.01 \text{ g C}}{44.01 \text{ g}}$ × 100 = 27.29% C

    % O = $\dfrac{32.00 \text{ g O}}{44.01 \text{ g}}$ × 100 = 72.71% O

h. mass of Ag present =   107.9 g =   107.9 g
    mass of N present =   14.01 g =   14.01 g
    mass of O present =   3(16.00 g) =   48.00 g
    ──
    molar mass of $AgNO_3$ =   169.9 g

    % Ag = $\dfrac{107.9 \text{ g Ag}}{169.9 \text{ g}}$ × 100 = 63.51% Ag

    % N = $\dfrac{14.01 \text{ g N}}{169.9 \text{ g}}$ × 100 = 8.246% N

    % O = $\dfrac{48.00 \text{ g O}}{169.9 \text{ g}}$ × 100 = 28.25% O

120.
a. % Fe = $\dfrac{55.85 \text{ g Fe}}{151.92 \text{ g}}$ × 100 = 36.76% Fe

b. % Ag = $\dfrac{215.8 \text{ g Ag}}{231.8 \text{ g}}$ × 100 = 93.10% Ag

c. % Sr = $\dfrac{87.62 \text{ g Sr}}{158.5 \text{ g}}$ × 100 = 55.28% Sr

d. % C = $\dfrac{48.04 \text{ g C}}{86.09 \text{ g}}$ × 100 = 55.80% C

e. % C = $\dfrac{12.01 \text{ g C}}{32.04 \text{ g}}$ × 100 = 37.48% C

f. % Al = $\dfrac{53.96 \text{ g Al}}{101.96 \text{ g}}$ × 100 = 52.92% Al

g. % K = $\dfrac{39.10 \text{ g K}}{106.55 \text{ g}}$ × 100 = 36.70% K

h. % K = $\dfrac{39.10 \text{ g K}}{74.55 \text{ g}}$ × 100 = 52.45% K

96 Chapter 8 Chemical Composition

122. $0.2990 \text{ g C} \times \dfrac{1 \text{ mol C}}{12.01 \text{ g C}} = 0.02490 \text{ mol C}$

$0.05849 \text{ g H} \times \dfrac{1 \text{ mol H}}{1.008 \text{ g H}} = 0.05803 \text{ mol H}$

$0.2318 \text{ g N} \times \dfrac{1 \text{ mol N}}{14.01 \text{ g N}} = 0.01655 \text{ mol N}$

$0.1328 \text{ g O} \times \dfrac{1 \text{ mol O}}{16.00 \text{ g O}} = 0.008300 \text{ mol O}$

Dividing each number of moles by the smallest number of moles (0.008300 mol O) gives

$\dfrac{0.02490 \text{ mol C}}{0.008300} = 3.000 \text{ mol C}$

$\dfrac{0.05803 \text{ mol H}}{0.008300} = 6.992 \text{ mol H}$

$\dfrac{0.01655 \text{ mol N}}{0.008300} = 1.994 \text{ mol N}$

$\dfrac{0.008300 \text{ mol O}}{0.008300} = 1.000 \text{ mol O}$

The empirical formula is $C_3H_7N_2O$.

124. Mass of oxygen in compound = 4.33 g − 4.01 g = 0.32 g O

$4.01 \text{ g Hg} \times \dfrac{1 \text{ mol Hg}}{200.6 \text{ g Hg}} = 0.0200 \text{ mol Hg}$

$0.32 \text{ g O} \times \dfrac{1 \text{ mol O}}{16.00 \text{ g O}} = 0.020 \text{ mol O}$

Since the numbers of moles are equal, the empirical formula is HgO.

126. Assume we have 100.0 g of the compound.

$65.95 \text{ g Ba} \times \dfrac{1 \text{ mol Ba}}{137.3 \text{ g Ba}} = 0.4803 \text{ mol Ba}$

$34.05 \text{ g Cl} \times \dfrac{1 \text{ mol Cl}}{35.45 \text{ g Cl}} = 0.9605 \text{ mol Cl}$

Dividing each of these number of moles by the smaller number gives

$\dfrac{0.4803 \text{ mol Ba}}{0.4803} = 1.000 \text{ mol Ba}$

$$\frac{0.9605 \text{ mol Cl}}{0.4803} = 2.000 \text{ mol Cl}$$

The empirical formula is then $BaCl_2$.

Chapter 9   Chemical Quantities

2. The coefficients of the balanced chemical equation for a reaction indicate the *relative numbers of moles* of each reactant that combine during the process, as well as the number of moles of each product formed.

4. Balanced chemical equations tell us in what proportions *on a mole basis* substances combine; since the molar masses of $C(s)$ and $O_2(g)$ are different, 1 g of O_2 could not represent the same number of moles as 1 g of C.

6. a. $3MnO_2(s) + 4Al(s) \rightarrow 3Mn(s) + 2Al_2O_3(s)$

    Three formula units of manganese(IV) oxide react with four aluminum atoms, producing three manganese atoms and two formula units of aluminum oxide. Three moles of solid manganese(IV) oxide react with four moles of solid aluminum, to produce three moles of solid manganese and two moles of solid aluminum oxide.

    b. $B_2O_3(s) + 3CaF_2(s) \rightarrow 2BF_3(g) + 3CaO(s)$

    One molecule of diboron trioxide reacts with three formula units of calcium fluoride, producing two molecules of boron trifluoride and three formula units of calcium oxide. One mole of solid diboron trioxide reacts with three moles of solid calcium fluoride, to give two moles of gaseous boron trifluoride and three moles of solid calcium oxide.

    c. $3NO_2(g) + H_2O(l) \rightarrow 2HNO_3(aq) + NO(g)$

    Three molecules of nitrogen dioxide [nitrogen(IV) oxide] react with one molecule of water, to produce two molecules of nitric acid and one molecule of nitrogen monoxide [nitrogen(II) oxide]. Three moles of gaseous nitrogen dioxide react with one mole of liquid water, to produce two moles of aqueous nitric acid and one mole of nitrogen monoxide gas.

    d. $C_6H_6(g) + 3H_2(g) \rightarrow C_6H_{12}(g)$

    One molecule of C_6H_6 (which is named benzene) reacts with three molecules of hydrogen, producing just one molecule of C_6H_{12} (which is named cyclohexane). One mole of gaseous benzene reacts with three moles of hydrogen gas, giving one mole of gaseous cyclohexane.

8. False. Reactions take place on a *mole* basis: one *mole* of nitrogen gas would react with three *moles* of iodine, giving two *moles* of nitrogen triiodide.

10. $2Ag(s) + H_2S(g) \rightarrow Ag_2S(s) + H_2(g)$

$$\frac{1 \text{ mol } Ag_2S}{2 \text{ mol } Ag} \quad \text{and} \quad \frac{1 \text{ mol } H_2}{2 \text{ mol } Ag}$$

12. a. $2FeO(s) + C(s) \rightarrow 2Fe(l) + CO_2(g)$

$$0.125 \text{ mol FeO} \times \frac{2 \text{ mol Fe}}{2 \text{ mol FeO}} = 0.125 \text{ mol Fe}$$

$$0.125 \text{ mol FeO} \times \frac{1 \text{ mol } CO_2}{2 \text{ mol FeO}} = 0.0625 \text{ mol } CO_2$$

b. $Cl_2(g) + 2KI(aq) \rightarrow 2KCl(aq) + I_2(s)$

$$0.125 \text{ mol KI} \times \frac{2 \text{ mol KCl}}{2 \text{ mol KI}} = 0.125 \text{ mol KCl}$$

$$0.125 \text{ mol KI} \times \frac{1 \text{ mol } I_2}{2 \text{ mol KI}} = 0.0625 \text{ mol } I_2$$

c. $Na_2B_4O_7(s) + H_2SO_4(aq) + 5H_2O(l) \rightarrow 4H_3BO_3(s) + Na_2SO_4(aq)$

$$0.125 \text{ mol } Na_2B_4O_7 \times \frac{4 \text{ mol } H_3BO_3}{1 \text{ mol } Na_2B_4O_7} = 0.500 \text{ mol } H_3BO_3$$

$$0.125 \text{ mol } Na_2B_4O_7 \times \frac{1 \text{ mol } Na_2SO_4}{1 \text{ mol } Na_2B_4O_7} = 0.125 \text{ mol } Na_2SO_4$$

d. $CaC_2(s) + 2H_2O(l) \rightarrow Ca(OH)_2(s) + C_2H_2(g)$

$$0.125 \text{ mol } CaC_2 \times \frac{1 \text{ mol } Ca(OH)_2}{1 \text{ mol } CaC_2} = 0.125 \text{ mol } Ca(OH)_2$$

$$0.125 \text{ mol } CaC_2 \times \frac{1 \text{ mol } C_2H_2}{1 \text{ mol } CaC_2} = 0.125 \text{ mol } C_2H_2$$

14. a. $NH_3(g) + HCl(g) \rightarrow NH_4Cl(s)$

molar mass of NH_4Cl, 53.49 g

$$0.50 \text{ mol } NH_3 \times \frac{1 \text{ mol } NH_4Cl}{1 \text{ mol } NH_3} = 0.50 \text{ mol } NH_4Cl$$

$$0.50 \text{ mol } NH_4Cl \times \frac{53.49 \text{ g } NH_4Cl}{1 \text{ mol } NH_4Cl} = 27 \text{ g } NH_4Cl$$

b. $CH_4(g) + 4S(g) \rightarrow CS_2(l) + 2H_2S(g)$

molar masses: CS_2, 76.15 g; H_2S, 34.09 g

$$0.50 \text{ mol S} \times \frac{1 \text{ mol } CS_2}{4 \text{ mol S}} = 0.125 \text{ mol } CS_2 \ (= 0.13 \text{ mol } CS_2)$$

100 Chapter 9   Chemical Quantities

$$0.125 \text{ mol CS}_2 \times \frac{76.15 \text{ g CS}_2}{1 \text{ mol CS}_2} = 9.5 \text{ g CS}_2$$

$$0.50 \text{ mol S} \times \frac{2 \text{ mol H}_2\text{S}}{4 \text{ mol S}} = 0.25 \text{ mol H}_2\text{S}$$

$$0.25 \text{ mol H}_2\text{S} \times \frac{34.09 \text{ g H}_2\text{S}}{1 \text{ mol H}_2\text{S}} = 8.5 \text{ g H}_2\text{S}$$

c. $PCl_3(l) + 3H_2O(l) \rightarrow H_3PO_3(aq) + 3HCl(aq)$

molar masses: H_3PO_3, 81.99 g; HCl, 36.46 g

$$0.50 \text{ mol PCl}_3 \times \frac{1 \text{ mol H}_3\text{PO}_3}{1 \text{ mol PCl}_3} = 0.50 \text{ mol H}_3\text{PO}_3$$

$$0.50 \text{ mol H}_3\text{PO}_3 \times \frac{81.99 \text{ g H}_3\text{PO}_3}{1 \text{ mol H}_3\text{PO}_3} = 41 \text{ g H}_3\text{PO}_3$$

$$0.50 \text{ mol PCl}_3 \times \frac{3 \text{ mol HCl}}{1 \text{ mol PCl}_3} = 1.5 \text{ mol HCl}$$

$$1.5 \text{ mol HCl} \times \frac{36.46 \text{ g HCl}}{1 \text{ mol HCl}} = 54.7 = 55 \text{ g HCl}$$

d. $NaOH(s) + CO_2(g) \rightarrow NaHCO_3(s)$

molar mass of $NaHCO_3$ = 84.01 g

$$0.50 \text{ mol NaOH} \times \frac{1 \text{ mol NaHCO}_3}{1 \text{ mol NaOH}} = 0.50 \text{ mol NaHCO}_3$$

$$0.50 \text{ mol NaHCO}_3 \times \frac{84.01 \text{ g NaHCO}_3}{1 \text{ mol NaHCO}_3} = 42 \text{ g NaHCO}_3$$

16. Before doing the calculations, the equations must be *balanced*.

a. $4KO_2(s) + 2H_2O(l) \rightarrow 3O_2(g) + 4KOH(s)$

$$0.625 \text{ mol KOH} \times \frac{3 \text{ mol O}_2}{4 \text{ mol KOH}} = 0.469 \text{ mol O}_2$$

b. $SeO_2(g) + 2H_2Se(g) \rightarrow 3Se(s) + 2H_2O(g)$

$$0.625 \text{ mol H}_2\text{O} \times \frac{3 \text{ mol Se}}{2 \text{ mol H}_2\text{O}} = 0.938 \text{ mol Se}$$

c. $2CH_3CH_2OH(l) + O_2(g) \rightarrow 2CH_3CHO(aq) + 2H_2O(l)$

$$0.625 \text{ mol H}_2\text{O} \times \frac{2 \text{ mol CH}_3\text{CHO}}{2 \text{ mol H}_2\text{O}} = 0.625 \text{ mol CH}_3\text{CHO}$$

d. $Fe_2O_3(s) + 2Al(s) \rightarrow 2Fe(l) + Al_2O_3(s)$

$$0.625 \text{ mol Al}_2\text{O}_3 \times \frac{2 \text{ mol Fe}}{1 \text{ mol Al}_2\text{O}_3} = 1.25 \text{ mol Fe}$$

18. Stoichiometry is the process of using a chemical equation to calculate the relative masses of reactants and products involved in a reaction.

20. a. molar mass of Ag = 107.9 g

$$2.01 \times 10^{-2} \text{ g Ag} \times \frac{1 \text{ mol}}{107.9 \text{ g}} = 1.86 \times 10^{-4} \text{ mol Ag}$$

b. molar mass of $(NH_4)_2S$ = 68.15 g

$$45.2 \text{ mg } (NH_4)_2S \times \frac{1 \text{ g}}{1000 \text{ mg}} \times \frac{1 \text{ mol}}{68.15 \text{ g}} = 6.63 \times 10^{-4} \text{ mol } (NH_4)_2S$$

c. molar mass of uranium = 238.0 g

$$61.7 \text{ } \mu\text{g U} \times \frac{1 \text{ g}}{10^6 \text{ } \mu\text{g}} \times \frac{1 \text{ mol}}{238.0 \text{ g}} = 2.59 \times 10^{-7} \text{ mol U}$$

d. molar mass of SO_2 = 64.07 g

$$5.23 \text{ kg } SO_2 \times \frac{1000 \text{ g}}{1 \text{ kg}} \times \frac{1 \text{ mol}}{64.07 \text{ g}} = 81.6 \text{ mol } SO_2$$

e. molar mass of $Fe(NO_3)_3$ = 241.9 g

$$272 \text{ g } Fe(NO_3)_3 \times \frac{1 \text{ mol}}{241.9 \text{ g}} = 1.12 \text{ mol } Fe(NO_3)_3$$

f. molar mass of $FeSO_4$ = 151.9 g

$$12.7 \text{ mg } FeSO_4 \times \frac{1 \text{ g}}{1000 \text{ mg}} \times \frac{1 \text{ mol}}{151.9 \text{ g}} = 8.36 \times 10^{-5} \text{ mol } FeSO_4$$

g. molar mass of LiOH = 23.95 g

$$6.91 \times 10^3 \text{ g LiOH} \times \frac{1 \text{ mol}}{23.95 \text{ g}} = 288.5 = 289 \text{ mol LiOH}$$

22. a. molar mass of $CaCO_3$ = 100.1 g

$$2.21 \times 10^{-4} \text{ mol } CaCO_3 \times \frac{100.1 \text{ g}}{1 \text{ mol}} = 0.0221 \text{ g } CaCO_3$$

b. molar mass of He = 4.003 g

$$2.75 \text{ mol He} \times \frac{4.003 \text{ g}}{1 \text{ mol}} = 11.0 \text{ g He}$$

c. molar mass of O_2 = 32.00 g

$$0.00975 \text{ mol } O_2 \times \frac{32.00 \text{ g}}{1 \text{ mol}} = 0.312 \text{ g } O_2$$

d. molar mass of CO_2 = 44.01 g

7.21 millimol = 0.00721 mol

102   Chapter 9   Chemical Quantities

$$0.00721 \text{ mol} \times \frac{44.01 \text{ g}}{1 \text{ mol}} = 0.317 \text{ g } CO_2$$

e. molar mass of FeS = 87.92 g

$$0.835 \text{ mol FeS} \times \frac{87.92 \text{ g}}{1 \text{ mol}} = 73.4 \text{ g FeS}$$

f. molar mass of KOH = 56.11 g

$$4.01 \text{ mol KOH} \times \frac{56.11 \text{ g}}{1 \text{ mol}} = 225 \text{ g KOH}$$

g. molar mass of H_2 = 2.016 g

$$0.0219 \text{ mol } H_2 \times \frac{2.016 \text{ g}}{1 \text{ mol}} = 0.0442 \text{ g } H_2$$

24. Before any calculations are done, the equations must be *balanced*.

a. $Mg(s) + CuCl_2(aq) \rightarrow MgCl_2(aq) + Cu(s)$

molar mass of Mg = 24.31 g

$$25.0 \text{ g Mg} \times \frac{1 \text{ mol}}{24.31 \text{ g}} = 1.03 \text{ mol Mg}$$

$$1.03 \text{ mol Mg} \times \frac{1 \text{ mol } CuCl_2}{1 \text{ mol Mg}} = 1.03 \text{ mol } CuCl_2$$

b. $2AgNO_3(aq) + NiCl_2(aq) \rightarrow 2AgCl(s) + Ni(NO_3)_2(aq)$

molar mass of $AgNO_3$ = 169.9 g

$$25.0 \text{ g } AgNO_3 \times \frac{1 \text{ mol}}{169.9 \text{ g}} = 0.147 \text{ mol } AgNO_3$$

$$0.147 \text{ mol } AgNO_3 \times \frac{1 \text{ mol } NiCl_2}{2 \text{ mol } AgNO_3} = 0.0735 \text{ mol } NiCl_2$$

c. $NaHSO_3(aq) + NaOH(aq) \rightarrow Na_2SO_3(aq) + H_2O(l)$

molar mass of $NaHSO_3$ = 104.1 g

$$25.0 \text{ g } NaHSO_3 \times \frac{1 \text{ mol}}{104.1 \text{ g}} = 0.240 \text{ mol } NaHSO_3$$

$$0.240 \text{ mol } NaHSO_3 \times \frac{1 \text{ mol NaOH}}{1 \text{ mol } NaHSO_3} = 0.240 \text{ mol NaOH}$$

d. $KHCO_3(aq) + HCl(aq) \rightarrow KCl(aq) + H_2O(l) + CO_2(g)$

molar mass of $KHCO_3$ = 100.1 g

$$25.0 \text{ g } KHCO_3 \times \frac{1 \text{ mol}}{100.1 \text{ g}} = 0.250 \text{ mol } KHCO_3$$

$$0.250 \text{ mol KHCO}_3 \times \frac{1 \text{ mol HCl}}{1 \text{ mol KHCO}_3} = 0.250 \text{ mol HCl}$$

26. Before any calculations are done, the equations must be *balanced*. Since the given and required quantities in this question are given in *milligrams*, it is most convenient to perform the calculations in terms of *millimoles* of the substances involved. One millimole of a substance represents the molar mass of the substance expressed in milligrams.

   a. $FeSO_4(aq) + K_2CO_3(aq) \rightarrow FeCO_3(s) + K_2SO_4(aq)$

   millimolar masses: $FeSO_4$, 151.9 mg; $FeCO_3$, 115.9 mg; K_2SO_4, 174.3 mg

   $$10.0 \text{ mg FeSO}_4 \times \frac{1 \text{ mmol FeSO}_4}{151.9 \text{ mg FeSO}_4} = 0.0658 \text{ mmol FeSO}_4$$

   $$0.0658 \text{ mmol FeSO}_4 \times \frac{1 \text{ mmol FeCO}_3}{1 \text{ mmol FeSO}_4} \times \frac{115.9 \text{ mg FeCO}_3}{1 \text{ mmol FeCO}_3} = 7.63 \text{ mg FeCO}_3$$

   $$0.0658 \text{ mmol FeSO}_4 \times \frac{1 \text{ mmol K}_2SO_4}{1 \text{ mmol FeSO}_4} \times \frac{174.3 \text{ mg K}_2SO_4}{1 \text{ mmol K}_2SO_4} = 11.5 \text{ mg K}_2SO_4$$

   b. $4Cr(s) + 3SnCl_4(l) \rightarrow 4CrCl_3(s) + 3Sn(s)$

   millimolar masses: Cr, 52.00 mg; $CrCl_3$, 158.4 mg; Sn, 118.7 mg

   $$10.0 \text{ mg Cr} \times \frac{1 \text{ mmol Cr}}{52.00 \text{ mg Cr}} = 0.192 \text{ mmol Cr}$$

   $$0.192 \text{ mmol Cr} \times \frac{4 \text{ mmol CrCl}_3}{4 \text{ mmol Cr}} \times \frac{158.4 \text{ mg CrCl}_3}{1 \text{ mmol CrCl}_3} = 30.4 \text{ mg CrCl}_3$$

   $$0.192 \text{ mmol Cr} \times \frac{3 \text{ mmol Sn}}{4 \text{ mmol Cr}} \times \frac{118.7 \text{ mg Sn}}{1 \text{ mmol Sn}} = 17.1 \text{ mg Sn}$$

   c. $16Fe(s) + 3S_8(s) \rightarrow 8Fe_2S_3(s)$

   millimolar masses: S_8, 256.6 mg; Fe_2S_3, 207.9 mg

   $$10.0 \text{ mg S}_8 \times \frac{1 \text{ mmol S}_8}{256.6 \text{ mg S}_8} = 0.0390 \text{ mmol S}_8$$

   $$0.0390 \text{ mmol S}_8 \times \frac{8 \text{ mmol Fe}_2S_3}{3 \text{ mmol S}_8} \times \frac{207.9 \text{ mg Fe}_2S_3}{1 \text{ mmol Fe}_2S_3} = 21.6 \text{ mg Fe}_2S_3$$

   d. $3Ag(s) + 4HNO_3(aq) \rightarrow 3AgNO_3(aq) + 2H_2O(l) + NO(g)$

   millimolar masses: HNO_3, 63.0 mg; $AgNO_3$, 169.9 mg
   H_2O, 18.0 mg; NO, 30.0 mg

   $$10.0 \text{ mg HNO}_3 \times \frac{1 \text{ mmol HNO}_3}{63.0 \text{ mg HNO}_3} = 0.159 \text{ mmol HNO}_3$$

   $$0.159 \text{ mmol HNO}_3 \times \frac{3 \text{ mmol AgNO}_3}{4 \text{ mmol HNO}_3} \times \frac{169.9 \text{ mg AgNO}_3}{1 \text{ mmol AgNO}_3} = 20.3 \text{ mg AgNO}_3$$

$$0.159 \text{ mmol HNO}_3 \times \frac{2 \text{ mmol H}_2\text{O}}{4 \text{ mmol HNO}_3} \times \frac{18.0 \text{ mg H}_2\text{O}}{1 \text{ mmol H}_2\text{O}} = 1.43 \text{ mg H}_2\text{O}$$

$$0.159 \text{ mmol HNO}_3 \times \frac{1 \text{ mmol NO}}{4 \text{ mmol HNO}_3} \times \frac{30.0 \text{ mg NO}}{1 \text{ mmol NO}} = 1.19 \text{ mg NO}$$

28. $2H_2(g) + O_2(g) \to 2H_2O(g)$

    molar masses of O_2 = 32.00 g

    $$0.0275 \text{ mol H}_2 \times \frac{1 \text{ mol O}_2}{2 \text{ mol H}_2} = 0.01375 = 0.0138 \text{ mol O}_2$$

    $$0.01375 \text{ mol O}_2 \times \frac{32.00 \text{ g O}_2}{1 \text{ mol O}_2} = 0.440 \text{ g O}_2$$

30. $Cl_2(g) + 2KI(aq) \to I_2(s) + 2KCl(aq)$

    molar masses: Cl_2, 70.90 g; I_2, 253.8 g

    $$2.55 \text{ g Cl}_2 \times \frac{1 \text{ mol Cl}_2}{70.90 \text{ g Cl}_2} = 0.0360 \text{ mol Cl}_2$$

    $$0.0360 \text{ mol Cl}_2 \times \frac{1 \text{ mol I}_2}{1 \text{ mol Cl}_2} = 0.0360 \text{ mol I}_2$$

    $$0.0360 \text{ mol I}_2 \times \frac{253.8 \text{ g I}_2}{1 \text{ mol I}_2} = 9.13 \text{ g I}_2$$

32. $C_2H_5OH(l) + 3O_2(g) \to 2CO_2(g) + 3H_2O(l)$

    molar masses: C_2H_5OH, 46.07 g; CO_2, 44.01 g

    $$25.0 \text{ g C}_2\text{H}_5\text{OH} \times \frac{1 \text{ mol}}{46.07 \text{ g}} = 0.543 \text{ mol C}_2\text{H}_5\text{OH}$$

    $$0.543 \text{ mol C}_2\text{H}_5\text{OH} \times \frac{2 \text{ mol CO}_2}{1 \text{ mol C}_2\text{H}_5\text{OH}} = 1.086 \text{ mol CO}_2$$

    $$1.086 \text{ mol CO}_2 \times \frac{44.01 \text{ g}}{1 \text{ mol}} = 47.8 \text{ g CO}_2$$

34. $Cl_2(g) + F_2(g) \to 2ClF(g)$

    molar masses: Cl_2, 70.90 g; F_2, 38.00 g

    $$5.00 \text{ mg F}_2 \times \frac{1 \text{ g}}{10^3 \text{ mg}} \times \frac{1 \text{ mol F}_2}{38.00 \text{ g F}_2} = 1.316 \times 10^{-4} \text{ mol F}_2$$

    $$1.316 \times 10^{-4} \text{ mol F}_2 \times \frac{1 \text{ mol Cl}_2}{1 \text{ mol F}_2} \times \frac{70.90 \text{ g Cl}_2}{1 \text{ mol Cl}_2} = 9.33 \times 10^{-3} \text{ g Cl}_2$$

36. $Cu(s) + S(s) \to CuS(s)$

    molar masses: Cu, 63.55 g; S, 32.07 g

$$1.25 \text{ g Cu} \times \frac{1 \text{ mol}}{63.55 \text{ g}} = 1.97 \times 10^{-2} \text{ mol Cu}$$

$$1.97 \times 10^{-2} \text{ mol Cu} \times \frac{1 \text{ mol S}}{1 \text{ mol Cu}} = 1.97 \times 10^{-2} \text{ mol S}$$

$$1.97 \times 10^{-2} \text{ mol S} \times \frac{32.07 \text{ g}}{1 \text{ mol}} = 0.631 \text{ g S}$$

38. $2\text{NaOH}(aq) + \text{CO}_2(g) \rightarrow \text{Na}_2\text{CO}_3(aq) + \text{H}_2\text{O}(l)$

    molar masses: NaOH, 40.00 g; CO_2, 44.01 g

    $$5.00 \text{ g NaOH} \times \frac{1 \text{ mol}}{40.00 \text{ g}} = 0.125 \text{ mol NaOH}$$

    $$0.125 \text{ mol NaOH} \times \frac{1 \text{ mol CO}_2}{2 \text{ mol NaOH}} = 6.25 \times 10^{-2} \text{ mol CO}_2$$

    $$6.25 \times 10^{-2} \text{ mol CO}_2 \times \frac{44.01 \text{ g}}{1 \text{ mol}} = 2.75 \text{ g CO}_2$$

40. $2\text{Mg}(s) + \text{O}_2(g) \rightarrow 2\text{MgO}(s)$

    molar masses: Mg, 24.31 g; MgO, 40.31 g

    $$1.25 \text{ g Mg} \times \frac{1 \text{ mol}}{24.31 \text{ g}} = 5.14 \times 10^{-2} \text{ mol Mg}$$

    $$5.14 \times 10^{-2} \text{ mol Mg} \times \frac{2 \text{ mol MgO}}{2 \text{ mol Mg}} = 5.14 \times 10^{-2} \text{ mol MgO}$$

    $$5.14 \times 10^{-2} \text{ mol MgO} \times \frac{40.31 \text{ g}}{1 \text{ mol}} = 2.07 \text{ g MgO}$$

42. To determine the limiting reactant, first calculate the number of moles of each reactant present. Then determine how these numbers of moles correspond to the stoichiometric ratio indicated by the balanced chemical equation for the reaction.

44. A reactant is present *in excess* if there is more of that reactant present than is needed to combine with the limiting reactant for the process. By definition, the limiting reactant cannot be present in excess. An excess of any reactant does not affect the theoretical yield for a process: the theoretical yield is determined by the limiting reactant.

46. a. $2\text{Al}(s) + 6\text{HCl}(aq) \rightarrow 2\text{AlCl}_3(aq) + 3\text{H}_2(g)$

    Molar masses: Al, 26.98 g; HCl, 36.46 g; AlCl_3, 133.3 g; H_2, 2.016 g

    $$15.0 \text{ g Al} \times \frac{1 \text{ mol}}{26.98 \text{ g}} = 0.556 \text{ mol Al}$$

$$15.0 \text{ g HCl} \times \frac{1 \text{ mol}}{36.46 \text{ g}} = 0.411 \text{ mol HCl}$$

Since HCl is needed to react with Al in a 6:2 (i.e., 3:1) molar ratio, it seems pretty certain that HCl is the limiting reactant. To prove this, we can calculate the quantity of Al that would react with the given number of moles of HCl.

$$0.411 \text{ mol HCl} \times \frac{2 \text{ mol Al}}{6 \text{ mol HCl}} = 0.137 \text{ mol Al}$$

By this calculation we have shown that *all* the HCl present will be needed to react with only 0.137 mol Al (out of the 0.556 mol Al present). Therefore HCl is the limiting reactant, and Al is present in excess. The calculation of the masses of products produced is based on the number of moles of the limiting reactant.

$$0.411 \text{ mol HCl} \times \frac{2 \text{ mol AlCl}_3}{6 \text{ mol HCl}} \times \frac{133.3 \text{ g}}{1 \text{ mol}} = 18.3 \text{ g AlCl}_3$$

$$0.411 \text{ mol HCl} \times \frac{3 \text{ mol H}_2}{6 \text{ mol HCl}} \times \frac{2.016 \text{ g}}{1 \text{ mol}} = 0.414 \text{ g H}_2$$

b. $2\text{NaOH}(aq) + \text{CO}_2(g) \rightarrow \text{Na}_2\text{CO}_3(aq) + \text{H}_2\text{O}(l)$

molar masses: NaOH, 40.00 g; CO_2, 44.01 g; Na_2CO_3, 105.99 g; H_2O, 18.02 g

$$15.0 \text{ g NaOH} \times \frac{1 \text{ mol}}{40.00 \text{ g}} = 0.375 \text{ mol NaOH}$$

$$15.0 \text{ g CO}_2 \times \frac{1 \text{ mol}}{44.01 \text{ g}} = 0.341 \text{ mol CO}_2$$

For the 0.375 mol NaOH, let's calculate if there is enough CO_2 present to react:

$$0.375 \text{ mol NaOH} \times \frac{1 \text{ mol CO}_2}{2 \text{ mol NaOH}} = 0.1875 \text{ mol CO}_2 \ (0.188 \text{ mol})$$

We have present *more* CO_2 (0.341 mol) than is needed to react with the given quantity of NaOH. NaOH is therefore the limiting reactant, and CO_2 is present in excess. The quantities of products resulting are based on the complete conversion of the limiting reactant (0.375 mol NaOH):

$$0.375 \text{ mol NaOH} \times \frac{1 \text{ mol Na}_2\text{CO}_3}{2 \text{ mol NaOH}} \times \frac{105.99 \text{ g}}{1 \text{ mol}} = 19.9 \text{ g Na}_2\text{CO}_3$$

$$0.375 \text{ mol NaOH} \times \frac{1 \text{ mol H}_2\text{O}}{2 \text{ mol NaOH}} \times \frac{18.02 \text{ g}}{1 \text{ mol}} = 3.38 \text{ g H}_2\text{O}$$

c. $Pb(NO_3)_2(aq) + 2HCl(aq) \rightarrow PbCl_2(s) + 2HNO_3(aq)$

Molar masses: $Pb(NO_3)_2$, 331.2 g; HCl, 36.46 g; $PbCl_2$, 278.1 g; HNO_3, 63.02 g

$$15.0 \text{ g } Pb(NO_3)_2 \times \frac{1 \text{ mol}}{331.2 \text{ g}} = 0.0453 \text{ mol } Pb(NO_3)_2$$

$$15.0 \text{ g HCl} \times \frac{1 \text{ mol}}{36.46 \text{ g}} = 0.411 \text{ mol}$$

With such a large disparity between the numbers of moles of the reactants, it's probably a sure bet that $Pb(NO_3)_2$ is the limiting reactant. To confirm this, we can calculate how many mol of HCl are needed to react with the given amount of $Pb(NO_3)_2$.

$$0.0453 \text{ mol } Pb(NO_3)_2 \times \frac{2 \text{ mol HCl}}{1 \text{ mol } Pb(NO_3)_2} = 0.0906 \text{ mol HCl}$$

We have considerably more HCl present than is needed to react completely with the $Pb(NO_3)_2$. Therefore $Pb(NO_3)_2$ is the limiting reactant, and HCl is present in excess. The quantities of products produced are based on the limiting reactant being completely consumed.

$$0.0453 \text{ mol } Pb(NO_3)_2 \times \frac{1 \text{ mol } PbCl_2}{1 \text{ mol } Pb(NO_3)_2} \times \frac{278.1 \text{ g}}{1 \text{ mol}} = 12.6 \text{ g } PbCl_2$$

$$0.0453 \text{ mol } Pb(NO_3)_2 \times \frac{2 \text{ mol } HNO_3}{1 \text{ mol } Pb(NO_3)_2} \times \frac{63.02 \text{ g}}{1 \text{ mol}} = 5.71 \text{ g } HNO_3$$

d. $2K(s) + I_2(s) \rightarrow 2KI(s)$

Molar masses: K, 39.10 g; I_2, 253.8 g; KI, 166.0 g

$$15.0 \text{ g K} \times \frac{1 \text{ mol}}{39.10 \text{ g}} = 0.384 \text{ mol K}$$

$$15.0 \text{ g } I_2 \times \frac{1 \text{ mol}}{253.8 \text{ g}} = 0.0591 \text{ mol } I_2$$

Since, from the balanced chemical equation, we need twice as many moles of K as moles of I_2, and since there is so little I_2 present, it's a safe bet that I_2 is the limiting reactant. To confirm this, we can calculate how many moles of K are needed to react with the given amount of I_2 present:

$$0.0591 \text{ mol } I_2 \times \frac{2 \text{ mol K}}{1 \text{ mol } I_2} = 0.1182 \text{ mol K}$$

Clearly we have more potassium present than is needed to react with the small amount of I_2 present. I_2 is therefore the limiting reactant, and potassium is present in excess. The amount of product

108   Chapter 9   Chemical Quantities

produced is calculated from the number of moles of the limiting reactant present:

$$0.0591 \text{ mol } I_2 \times \frac{2 \text{ mol KI}}{1 \text{ mol } I_2} \times \frac{166.0 \text{ g}}{1 \text{ mol}} = 19.6 \text{ g KI}$$

48. a. $2NH_3(g) + 2Na(s) \rightarrow 2NaNH_2(s) + H_2(g)$

Molar masses: NH_3, 17.03 g; Na, 22.99 g; $NaNH_2$, 39.02 g

$$50.0 \text{ g } NH_3 \times \frac{1 \text{ mol}}{17.03 \text{ g}} = 2.94 \text{ mol } NH_3$$

$$50.0 \text{ g Na} \times \frac{1 \text{ mol}}{22.99 \text{ g}} = 2.17 \text{ mol Na}$$

Since the coefficients of NH_3 and Na are the *same* in the balanced chemical equation for the reaction, the two reactants combine in a 1:1 molar ratio. Therefore Na is the limiting reactant, which will control the amount of product produced.

$$2.17 \text{ mol Na} \times \frac{2 \text{ mol } NaNH_2}{2 \text{ mol Na}} \times \frac{39.02 \text{ g}}{1 \text{ mol}} = 84.7 \text{ g } NaNH_2$$

b. $BaCl_2(aq) + Na_2SO_4(aq) \rightarrow BaSO_4(s) + 2NaCl(aq)$

Molar masses: $BaCl_2$, 208.2 g; Na_2SO_4, 142.1 g; $BaSO_4$, 233.4 g

$$50.0 \text{ g } BaCl_2 \times \frac{1 \text{ mol}}{208.2 \text{ g}} = 0.240 \text{ mol } BaCl_2$$

$$50.0 \text{ g } Na_2SO_4 \times \frac{1 \text{ mol}}{142.1 \text{ g}} = 0.352 \text{ mol } Na_2SO_4$$

Since the coefficients of $BaCl_2$ and Na_2SO_4 are the *same* in the balanced chemical equation for the reaction, the reactant having the smaller number of moles present ($BaCl_2$) must be the limiting reactant, which will control the amount of product produced.

$$0.240 \text{ mol } BaCl_2 \times \frac{1 \text{ mol } BaSO_4}{1 \text{ mol } BaCl_2} \times \frac{233.4 \text{ g}}{1 \text{ mol}} = 56.0 \text{ g } BaSO_4$$

c. $SO_2(g) + 2NaOH(aq) \rightarrow Na_2SO_3(aq) + H_2O(l)$

Molar masses: SO_2, 64.07 g; NaOH, 40.00 g; Na_2SO_3, 126.1 g

$$50.0 \text{ g } SO_2 \times \frac{1 \text{ mol}}{64.07 \text{ g}} = 0.780 \text{ mol } SO_2$$

$$50.0 \text{ g NaOH} \times \frac{1 \text{ mol}}{40.00 \text{ g}} = 1.25 \text{ mol NaOH}$$

From the balanced chemical equation for the reaction, every time one mol of SO_2 reacts, two mol of NaOH are needed. For 0.780 mol of SO_2,

2(0.780 mol) = 1.56 mol of NaOH would be needed. We do not have sufficient NaOH to react with the SO_2 present: therefore, NaOH is the limiting reactant, and controls the amount of product obtained.

$$1.25 \text{ mol NaOH} \times \frac{1 \text{ mol Na}_2\text{SO}_3}{2 \text{ mol NaOH}} \times \frac{126.1 \text{ g}}{1 \text{ mol}} = 78.8 \text{ g Na}_2\text{SO}_3$$

d. $2Al(s) + 3H_2SO_4(l) \rightarrow Al_2(SO_4)_3(s) + 3H_2(g)$

Molar masses: Al, 26.98 g; H_2SO_4, 98.09 g; $Al_2(SO_4)_3$, 342.2 g

$$50.0 \text{ g Al} \times \frac{1 \text{ mol}}{26.98 \text{ g}} = 1.85 \text{ mol Al}$$

$$50.0 \text{ g H}_2\text{SO}_4 \times \frac{1 \text{ mol}}{98.09 \text{ g}} = 0.510 \text{ mol H}_2\text{SO}_4$$

Since the amount of H_2SO_4 present is smaller than the amount of Al, let's see if H_2SO_4 is the limiting reactant, by calculating how much Al would react with the given amount of H_2SO_4.

$$0.510 \text{ mol H}_2\text{SO}_4 \times \frac{2 \text{ mol Al}}{3 \text{ mol H}_2\text{SO}_4} = 0.340 \text{ mol Al}$$

Since all the H_2SO_4 present would react with only a small portion of the Al present, H_2SO_4 is therefore the limiting reactant and will control the amount of product obtained.

$$0.510 \text{ mol H}_2\text{SO}_4 \times \frac{1 \text{ mol Al}_2(\text{SO}_4)_3}{3 \text{ mol H}_2\text{SO}_4} \times \frac{342.2 \text{ g}}{1 \text{ mol}} = 58.2 \text{ g Al}_2(\text{SO}_4)_3$$

50. a. $CO(g) + 2H_2(g) \rightarrow CH_3OH(l)$

CO is the limiting reactant; 11.4 mg CH_3OH

b. $2Al(s) + 3I_2(s) \rightarrow 2AlI_3(s)$

I_2 is the limiting reactant; 10.7 mg AlI_3

c. $Ca(OH)_2(aq) + 2HBr(aq) \rightarrow CaBr_2(aq) + 2H_2O(l)$

HBr is the limiting reactant; 12.4 mg $CaBr_2$; 2.23 mg H_2O

d. $2Cr(s) + 2H_3PO_4(aq) \rightarrow 2CrPO_4(s) + 3H_2(g)$

H_3PO_4 is the limiting reactant; 15.0 mg $CrPO_4$; 0.309 mg H_2

52. $2NH_3(g) + CO_2(g) \rightarrow CN_2H_4O(s) + H_2O(l)$

molar masses: NH_3, 17.03 g; CO_2, 44.01 g; CN_2H_4O, 60.06 g

110   Chapter 9   Chemical Quantities

$$100. \text{ g NH}_3 \times \frac{1 \text{ mol NH}_3}{17.03 \text{ g NH}_3} = 5.872 \text{ mol NH}_3$$

$$100. \text{ g CO}_2 \times \frac{1 \text{ mol CO}_2}{44.01 \text{ g CO}_2} = 2.272 \text{ mol CO}_2$$

See if CO_2 is the limiting reactant.

$$2.272 \text{ mol CO}_2 \times \frac{2 \text{ mol NH}_3}{1 \text{ mol CO}_2} = 4.544 \text{ mol NH}_3$$

CO_2 is indeed the limiting reactant.

$$2.272 \text{ mol CO}_2 \times \frac{1 \text{ mol CN}_2\text{H}_4\text{O}}{1 \text{ mol CO}_2} = 2.272 \text{ mol CN}_2\text{H}_4\text{O}$$

$$2.272 \text{ mol CN}_2\text{H}_4\text{O} \times \frac{60.06 \text{ g CN}_2\text{H}_4\text{O}}{1 \text{ mol CN}_2\text{H}_4\text{O}} = 136 \text{ g CN}_2\text{H}_4\text{O}$$

54. $4\text{Fe}(s) + 3\text{O}_2(g) \rightarrow 2\text{Fe}_2\text{O}_3(s)$

Molar masses: Fe, 55.85 g; Fe_2O_3, 159.7 g

$$1.25 \text{ g Fe} \times \frac{1 \text{ mol}}{55.85 \text{ g}} = 0.0224 \text{ mol Fe present}$$

Calculate how many mol of O_2 are required to react with this amount of Fe

$$0.0224 \text{ mol Fe} \times \frac{3 \text{ mol O}_2}{4 \text{ mol Fe}} = 0.0168 \text{ mol O}_2$$

Since we have more O_2 than this, Fe must be the limiting reactant.

$$0.0224 \text{ mol Fe} \times \frac{2 \text{ mol Fe}_2\text{O}_3}{4 \text{ mol Fe}} = 0.0112 \text{ mol Fe}_2\text{O}_3$$

$$0.0112 \text{ mol Fe}_2\text{O}_3 \times \frac{159.7 \text{ g Fe}_2\text{O}_3}{1 \text{ mol Fe}_2\text{O}_3} = 1.79 \text{ g Fe}_2\text{O}_3$$

56. $2\text{CuSO}_4(aq) + 5\text{KI}(aq) \rightarrow 2\text{CuI}(s) + \text{KI}_3(aq) + 2\text{K}_2\text{SO}_4(aq)$

Molar masses: $CuSO_4$, 159.6 g; KI, 166.0 g; CuI, 190.5 g; KI_3, 419.8 g; K_2SO_4, 174.3 g

$$0.525 \text{ g CuSO}_4 \times \frac{1 \text{ mol}}{159.6 \text{ g}} = 3.29 \times 10^{-3} \text{ mol CuSO}_4$$

$$2.00 \text{ g KI} \times \frac{1 \text{ mol}}{166.0 \text{ g}} = 0.0120 \text{ mol KI}$$

To determine the limiting reactant, let's calculate what amount of KI would be needed to react with the given amount of $CuSO_4$ present

$$3.29 \times 10^{-3} \text{ mol } CuSO_4 \times \frac{5 \text{ mol KI}}{2 \text{ mol } CuSO_4} = 8.23 \times 10^{-3} \text{ mol KI}$$

Since we have more KI present than the amount required to react with the $CuSO_4$ present, $CuSO_4$ must be the limiting reactant, which will control the amounts of products produced.

$$3.29 \times 10^{-3} \text{ mol } CuSO_4 \times \frac{2 \text{ mol CuI}}{2 \text{ mol } CuSO_4} \times \frac{190.5 \text{ g}}{1 \text{ mol}} = 0.627 \text{ g CuI}$$

$$3.29 \times 10^{-3} \text{ mol } CuSO_4 \times \frac{1 \text{ mol } KI_3}{2 \text{ mol } CuSO_4} \times \frac{419.8 \text{ g}}{1 \text{ mol}} = 0.691 \text{ g } KI_3$$

$$3.29 \times 10^{-3} \text{ mol } CuSO_4 \times \frac{2 \text{ mol } K_2SO_4}{2 \text{ mol } CuSO_4} \times \frac{174.3 \text{ g}}{1 \text{ mol}} = 0.573 \text{g } K_2SO_4$$

58. $SiO_2(s) + 3C(s) \rightarrow 2CO(g) + SiC(s)$

    molar masses: SiO_2, 60.09 g; SiC, 40.10 g

    $1.0 \text{ kg} = 1.0 \times 10^3 \text{ g}$

    $$1.0 \times 10^3 \text{ g } SiO_2 \times \frac{1 \text{ mol } SiO_2}{60.09 \text{ g } SiO_2} = 16.64 \text{ mol } SiO_2$$

    From the balanced chemical equation, if 16.64 mol of SiO_2 were to react completely (an excess of carbon is present), then 16.64 mol of SiC should be produced (the coefficients of SiO_2 and SiC are the same).

    $$16.64 \text{ mol SiC} \times \frac{40.01 \text{ g SiC}}{1 \text{ mol SiC}} = 6.7 \times 10^2 \text{ g SiC} = 0.67 \text{ kg SiC}$$

60. If the reaction is performed in a solvent, the product may have a substantial solubility in the solvent; the reaction may come to equilibrium before the full yield of product is achieved (see Chapter 16); loss of product may occur through operator error.

62. Percent yield = $\dfrac{\text{actual yield}}{\text{theoretical yield}} \times 100 = \dfrac{1.279 \text{ g}}{1.352 \text{ g}} \times 100 = 94.60\%$

64. $2LiOH(s) + CO_2(g) \rightarrow Li_2CO_3(s) + H_2O(g)$

    molar masses: LiOH, 23.95 g; CO_2, 44.01 g

    $$155 \text{ g LiOH} \times \frac{1 \text{ mol LiOH}}{23.95 \text{ g LiOH}} \times \frac{1 \text{ mol } CO_2}{2 \text{ mol LiOH}} \times \frac{44.01 \text{ g } CO_2}{1 \text{ mol } CO_2} = 142 \text{ g } CO_2$$

112 Chapter 9 Chemical Quantities

Since the cartridge has only absorbed 102 g CO_2 out of a total capacity of 142 g CO_2, the cartridge has absorbed

$$\frac{102 \text{ g}}{142 \text{ g}} \times 100 = 71.8\% \text{ of its capacity}$$

66. $CaCO_3(s) + 2HCl(g) \rightarrow CaCl_2(s) + CO_2(g) + H_2O(g)$

molar masses: $CaCO_3$, 100.1 g; HCl, 36.46 g; $CaCl_2$, 111.0 g

$$155 \text{ g CaCO}_3 \times \frac{1 \text{ mol CaCO}_3}{100.1 \text{ g CaCO}_3} = 1.548 \text{ mol CaCO}_3$$

$$250. \text{ g HCl} \times \frac{1 \text{ mol HCl}}{36.46 \text{ g HCl}} = 6.857 \text{ mol HCl}$$

$CaCO_3$ is the limiting reactant.

$$1.548 \text{ mol CaCO}_3 \times \frac{1 \text{ mol CaCl}_2}{1 \text{ mol CaCO}_3} = 1.548 \text{ mol CaCl}_2$$

$$1.548 \text{ mol CaCl}_2 \times \frac{111.0 \text{ g CaCl}_2}{1 \text{ mol CaCl}_2} = 172 \text{ g CaCl}_2$$

Percent yield = $\frac{\text{actual yield}}{\text{theoretical yield}} \times 100 = \frac{142 \text{ g}}{172 \text{ g}} \times 100 = 82.6\%$

68. $NaCl(aq) + NH_3(aq) + H_2O(l) + CO_2(s) \rightarrow NH_4Cl(aq) + NaHCO_3(s)$

molar masses: NH_3, 17.03 g; CO_2, 44.01 g; $NaHCO_3$, 84.01 g

$$10.0 \text{ g NH}_3 \times \frac{1 \text{ mol NH}_3}{17.03 \text{ g NH}_3} = 0.5872 \text{ mol NH}_3$$

$$15.0 \text{ g CO}_2 \times \frac{1 \text{ mol CO}_2}{44.01 \text{ g CO}_2} = 0.3408 \text{ mol CO}_2$$

CO_2 is the limiting reactant.

$$0.3408 \text{ mol CO}_2 \times \frac{1 \text{ mol NaHCO}_3}{1 \text{ mol CO}_2} = 0.3408 \text{ mol NaHCO}_3$$

$$0.3408 \text{ mol NaHCO}_3 \times \frac{84.01 \text{ g NaHCO}_3}{1 \text{ mol NaHCO}_3} = 28.6 \text{ g NaHCO}_3$$

70. $C_6H_{12}O_6(s) + 6O_2(g) \rightarrow 6CO_2(g) + 6H_2O(g)$

molar masses: glucose, 180.2 g; CO_2, 44.01 g

$$1.00 \text{ g glucose} \times \frac{1 \text{ mol glucose}}{180.2 \text{ g glucose}} = 5.549 \times 10^{-3} \text{ mol glucose}$$

$$5.549 \times 10^{-3} \text{ mol glucose} \times \frac{6 \text{ mol CO}_2}{1 \text{ mol glucose}} = 3.33 \times 10^{-2} \text{ mol CO}_2$$

$$3.33 \times 10^{-2} \text{ mol CO}_2 \times \frac{44.01 \text{ g CO}_2}{1 \text{ mol CO}_2} = 1.47 \text{ g CO}_2$$

72. $Ba^{2+}(aq) + SO_4^{2-}(aq) \rightarrow BaSO_4(s)$

    millimolar ionic masses: Ba^{2+}, 137.3 mg; SO_4^{2-}, 96.07 mg; $BaCl_2$, 208.2 mg

    $$150 \text{ mg SO}_4^{2-} \times \frac{1 \text{ mmol SO}_4^{2-}}{96.07 \text{ mg SO}_4^{2-}} = 1.56 \text{ millimol SO}_4^{2-}$$

    Since barium ion and sulfate ion react on a 1:1 stoichiometric basis, then 1.56 millimol of barium ion is needed, which corresponds to 1.56 millimol of $BaCl_2$

    $$1.56 \text{ millimol BaCl}_2 \times \frac{208.2 \text{ mg BaCl}_2}{1 \text{ millimol BaCl}_2} = 325 \text{ milligrams BaCl}_2 \text{ needed}$$

74. a. $UO_2(s) + 4HF(aq) \rightarrow UF_4(aq) + 2H_2O(l)$

    One molecule (formula unit) of uranium(IV) oxide will combine with four molecules of hydrofluoric acid, producing one uranium(IV) fluoride molecule and two water molecules. One mole of uranium(IV) oxide will combine with four moles of hydrofluoric acid to produce one mole of uranium(IV) fluoride and two moles of water.

    b. $2NaC_2H_3O_2(aq) + H_2SO_4(aq) \rightarrow Na_2SO_4(aq) + 2HC_2H_3O_2(aq)$

    Two molecules (formula units) of sodium acetate react exactly with one molecule of sulfuric acid, producing one molecule (formula unit) of sodium sulfate and two molecules of acetic acid. Two moles of sodium acetate will combine with one mole of sulfuric acid, producing one mole of sodium sulfate and two moles of acetic acid.

    c. $Mg(s) + 2HCl(aq) \rightarrow MgCl_2(aq) + H_2(g)$

    One magnesium atom will react with two hydrochloric acid molecules (formula units) to produce one molecule (formula unit) of magnesium chloride and one molecule of hydrogen gas. One mole of magnesium will combine with two moles of hydrochloric acid, producing one mole of magnesium chloride and one mole of gaseous hydrogen.

    d. $B_2O_3(s) + 3H_2O(l) \rightarrow 2B(OH)_3(aq)$

    One molecule of diboron trioxide will react exactly with three molecules of water, producing two molecules of boron trihydroxide (boric acid). One mole of diboron trioxide will combine with three moles of water to produce two moles of boron trihydroxide (boric acid).

76. For O_2: $\dfrac{5 \text{ mol } O_2}{1 \text{ mol } C_3H_8}$ For CO_2: $\dfrac{3 \text{ mol } CO_2}{1 \text{ mol } C_3H_8}$ For H_2O: $\dfrac{4 \text{ mol } H_2O}{1 \text{ mol } C_3H_8}$

114 Chapter 9 Chemical Quantities

78. a. $NH_3(g) + HCl(g) \rightarrow NH_4Cl(s)$

molar mass of NH_3 = 17.0 g

$$1.00 \text{ g } NH_3 \times \frac{1 \text{ mol } NH_3}{17.0 \text{ g } NH_3} = 0.0588 \text{ mol } NH_3$$

$$0.0588 \text{ mol } NH_3 \times \frac{1 \text{ mol } NH_4Cl}{1 \text{ mol } NH_3} = 0.0588 \text{ mol } NH_4Cl$$

b. $CaO(s) + CO_2(g) \rightarrow CaCO_3(s)$

molar mass CaO = 56.1 g

$$1.00 \text{ g CaO} \times \frac{1 \text{ mol CaO}}{56.1 \text{ g CaO}} = 0.0178 \text{ mol CaO}$$

$$0.0178 \text{ mol CaO} \times \frac{1 \text{ mol } CaCO_3}{1 \text{ mol CaO}} = 0.0178 \text{ mol } CaCO_3$$

c. $4Na(s) + O_2(g) \rightarrow 2Na_2O(s)$

molar mass Na = 22.99 g

$$1.00 \text{ g Na} \times \frac{1 \text{ mol Na}}{22.99 \text{ g Na}} = 0.0435 \text{ mol Na}$$

$$0.0435 \text{ mol Na} \times \frac{2 \text{ mol } Na_2O}{4 \text{ mol Na}} = 0.0217 \text{ mol } Na_2O$$

d. $2P(s) + 3Cl_2(g) \rightarrow 2PCl_3(l)$

molar mass P = 30.97 g

$$1.00 \text{ g P} \times \frac{1 \text{ mol P}}{30.97 \text{ g P}} = 0.0323 \text{ mol P}$$

$$0.0323 \text{ mol P} \times \frac{2 \text{ mol } PCl_3}{2 \text{ mol P}} = 0.0323 \text{ mol } PCl_3$$

80. a. molar mass HNO_3 = 63.0 g

$$5.0 \text{ mol } HNO_3 \times \frac{63.0 \text{ g } HNO_3}{1 \text{ mol } HNO_3} = 3.2 \times 10^2 \text{ g } HNO_3$$

b. molar mass Hg = 200.6 g

$$0.000305 \text{ mol Hg} \times \frac{200.6 \text{ g Hg}}{1 \text{ mol Hg}} = 0.0612 \text{ g Hg}$$

c. molar mass K_2CrO_4 = 194.2 g

$$2.31 \times 10^{-5} \text{ mol } K_2CrO_4 \times \frac{194.2 \text{ g } K_2CrO_4}{1 \text{ mol } K_2CrO_4} = 4.49 \times 10^{-3} \text{ g } K_2CrO_4$$

d. molar mass $AlCl_3$ = 133.3 g

$$10.5 \text{ mol } AlCl_3 \times \frac{133.3 \text{ g } AlCl_3}{1 \text{ mol } AlCl_3} = 1.40 \times 10^3 \text{ g } AlCl_3$$

e. molar mass SF_6 = 146.1 g

$$4.9 \times 10^4 \text{ mol } SF_6 \times \frac{146.1 \text{ g } SF_6}{1 \text{ mol } SF_6} = 7.2 \times 10^6 \text{ g } SF_6$$

f. molar mass NH_3 = 17.0 g

$$125 \text{ mol } NH_3 \times \frac{17.0 \text{ g } NH_3}{1 \text{ mol } NH_3} = 2.13 \times 10^3 \text{ g } NH_3$$

g. molar mass Na_2O_2 = 77.98 g

$$0.01205 \text{ mol } Na_2O_2 \times \frac{77.98 \text{ g } Na_2O_2}{1 \text{ mol } Na_2O_2} = 0.9397 \text{ g } Na_2O_2$$

82. $2SO_2(g) + O_2(g) \rightarrow 2SO_3(g)$

molar masses: SO_2, 64.07 g; SO_3, 80.07 g

150 kg = 1.5×10^5 g

$$1.5 \times 10^5 \text{ g } SO_2 \times \frac{1 \text{ mol } SO_2}{64.07 \text{ g } SO_2} = 2.34 \times 10^3 \text{ mol } SO_2$$

$$2.34 \times 10^3 \text{ mol } SO_2 \times \frac{2 \text{ mol } SO_3}{2 \text{ mol } SO_2} = 2.34 \times 10^3 \text{ mol } SO_3$$

$$2.34 \times 10^3 \text{ mol } SO_3 \times \frac{80.07 \text{ g } SO_3}{1 \text{ mol } SO_3} = 1.9 \times 10^5 \text{ g } SO_3 = 1.9 \times 10^2 \text{ kg } SO_3$$

84. $2Na_2O_2(s) + 2H_2O(l) \rightarrow 4NaOH(aq) + O_2(g)$

molar masses: Na_2O_2, 77.98 g; O_2, 32.00 g

$$3.25 \text{ g } Na_2O_2 \times \frac{1 \text{ mol } Na_2O_2}{77.98 \text{ g } Na_2O_2} = 0.0417 \text{ mol } Na_2O_2$$

$$0.0417 \text{ mol } Na_2O_2 \times \frac{1 \text{ mol } O_2}{2 \text{ mol } Na_2O_2} = 0.0209 \text{ mol } O_2$$

$$0.0209 \text{ mol } O_2 \times \frac{32.00 \text{ g } O_2}{1 \text{ mol } O_2} = 0.669 \text{ g } O_2$$

86. $Zn(s) + 2HCl(aq) \rightarrow ZnCl_2(aq) + H_2(g)$

molar masses: Zn, 65.38 g; H_2, 2.016 g

$$2.50 \text{ g Zn} \times \frac{1 \text{ mol Zn}}{65.38 \text{ g Zn}} = 0.03824 \text{ mol Zn}$$

$$0.03824 \text{ mol Zn} \times \frac{1 \text{ mol H}_2}{1 \text{ mol Zn}} = 0.03824 \text{ mol H}_2$$

$$0.03824 \text{ mol H}_2 \times \frac{2.016 \text{ g H}_2}{1 \text{ mol H}_2} = 0.0771 \text{ g H}_2$$

88. a. $2Na(s) + Br_2(l) \rightarrow 2NaBr(s)$

   molar masses: Na, 22.99 g; Br_2, 159.8 g; NaBr, 102.9 g

   $$5.0 \text{ g Na} \times \frac{1 \text{ mol Na}}{22.99 \text{ g Na}} = 0.2175 \text{ mol Na}$$

   $$5.0 \text{ g Br}_2 \times \frac{1 \text{ mol Br}_2}{159.8 \text{ g Br}_2} = 0.03129 \text{ mol Br}_2$$

   Intuitively, we would suspect that Br_2 is the limiting reactant, since there is much less Br_2 than Na on a mole basis. To *prove* that Br_2 is the limiting reactant, the following calculation is needed:

   $$0.03129 \text{ mol Br}_2 \times \frac{2 \text{ mol Na}}{1 \text{ mol Br}_2} = 0.06258 \text{ mol Na}$$

   Clearly, there is more Na than this present, so Br_2 limits the reaction extent and the amount of NaBr formed.

   $$0.03129 \text{ mol Br}_2 \times \frac{2 \text{ mol NaBr}}{1 \text{ mol Br}_2} = 0.06258 \text{ mol NaBr}$$

   $$0.06258 \text{ mol NaBr} \times \frac{102.9 \text{ g NaBr}}{1 \text{ mol NaBr}} = 6.4 \text{ g NaBr}$$

   b. $Zn(s) + CuSO_4(aq) \rightarrow ZnSO_4(aq) + Cu(s)$

   molar masses: Zn, 65.38 g; Cu, 63.55 g;
   $ZnSO_4$, 161.5 g; $CuSO_4$, 159.6 g

   $$5.0 \text{ g Zn} \times \frac{1 \text{ mol Zn}}{65.38 \text{ g Zn}} = 0.07648 \text{ mol Zn}$$

   $$5.0 \text{ g CuSO}_4 \times \frac{1 \text{ mol CuSO}_4}{159.6 \text{ g CuSO}_4} = 0.03132 \text{ mol CuSO}_4$$

   Since the coefficients of Zn and $CuSO_4$ are the *same* in the balanced chemical equation, an equal number of moles of Zn and $CuSO_4$ would be needed for complete reaction. Since there is less $CuSO_4$ present, $CuSO_4$ must clearly be the limiting reactant.

   $$0.03132 \text{ mol CuSO}_4 \times \frac{1 \text{ mol ZnSO}_4}{1 \text{ mol CuSO}_4} = 0.03132 \text{ mol ZnSO}_4$$

$$0.03132 \text{ mol ZnSO}_4 \times \frac{161.5 \text{ g ZnSO}_4}{1 \text{ mol ZnSO}_4} = 5.1 \text{ g ZnSO}_4$$

$$0.03132 \text{ mol CuSO}_4 \times \frac{1 \text{ mol Cu}}{1 \text{ mol CuSO}_4} = 0.03132 \text{ mol Cu}$$

$$0.03132 \text{ mol Cu} \times \frac{63.55 \text{ g Cu}}{1 \text{ mol Cu}} = 2.0 \text{ g Cu}$$

c. $NH_4Cl(aq) + NaOH(aq) \rightarrow NH_3(g) + H_2O(l) + NaCl(aq)$

molar masses: NH_4Cl, 53.49 g; NaOH, 40.00 g; NH_3, 17.03 g

H_2O, 18.02 g; NaCl, 58.44 g

$$5.0 \text{ g NH}_4\text{Cl} \times \frac{1 \text{ mol NH}_4\text{Cl}}{53.49 \text{ g NH}_4\text{Cl}} = 0.09348 \text{ mol NH}_4\text{Cl}$$

$$5.0 \text{ g NaOH} \times \frac{1 \text{ mol NaOH}}{40.00 \text{ g NaOH}} = 0.1250 \text{ mol NaOH}$$

Since the coefficients of NH_4Cl and NaOH are both *one* in the balanced chemical equation for the reaction, an equal number of moles of NH_4Cl and NaOH would be needed for complete reaction. Since there is less NH_4Cl present, NH_4Cl must be the limiting reactant.

Since the coefficients of the products in the balanced chemical equation are also all *one*, if 0.09348 mol of NH_4Cl (the limiting reactant) reacts completely, then 0.09348 mol of each product will be formed.

$$0.09348 \text{ mol NH}_3 \times \frac{17.03 \text{ g NH}_3}{1 \text{ mol NH}_3} = 1.6 \text{ g NH}_3$$

$$0.09348 \text{ mol H}_2\text{O} \times \frac{18.02 \text{ g H}_2\text{O}}{1 \text{ mol H}_2\text{O}} = 1.7 \text{ g H}_2\text{O}$$

$$0.09348 \text{ mol NaCl} \times \frac{58.44 \text{ g NaCl}}{1 \text{ mol NaCl}} = 5.5 \text{ g NaCl}$$

d. $Fe_2O_3(s) + 3CO(g) \rightarrow 2Fe(s) + 3CO_2(g)$

molar masses: Fe_2O_3, 159.7 g; CO, 28.01 g

Fe, 55.85 g; CO_2, 44.01 g

$$5.0 \text{ g Fe}_2\text{O}_3 \times \frac{1 \text{ mol Fe}_2\text{O}_3}{159.7 \text{ g Fe}_2\text{O}_3} = 0.03131 \text{ mol Fe}_2\text{O}_3$$

$$5.0 \text{ g CO} \times \frac{1 \text{ mol CO}}{28.01 \text{ g CO}} = 0.1785 \text{ mol CO}$$

Because there is considerably less Fe_2O_3 than CO on a mole basis, let's see if Fe_2O_3 is the limiting reactant.

$$0.03131 \text{ mol Fe}_2\text{O}_3 \times \frac{3 \text{ mol CO}}{1 \text{ mol Fe}_2\text{O}_3} = 0.09393 \text{ mol CO}$$

Since there is 0.1785 mol of CO present, but we have determined that only 0.09393 mol CO would be needed to react with all the Fe_2O_3 present, then Fe_2O_3 must be the limiting reactant. CO is present in excess.

$$0.03131 \text{ mol Fe}_2\text{O}_3 \times \frac{2 \text{ mol Fe}}{1 \text{ mol Fe}_2\text{O}_3} = 0.06262 \text{ mol Fe}$$

$$0.06262 \text{ mol Fe} \times \frac{55.85 \text{ g Fe}}{1 \text{ mol Fe}} = 3.5 \text{ g Fe}$$

$$0.03131 \text{ mol Fe}_2\text{O}_3 \times \frac{3 \text{ mol CO}_2}{1 \text{ mol Fe}_2\text{O}_3} = 0.09393 \text{ mol CO}_2$$

$$0.09393 \text{ mol CO}_2 \times \frac{44.01 \text{ g CO}_2}{1 \text{ mol CO}_2} = 4.1 \text{ g CO}_2$$

90. $N_2H_4(l) + O_2(g) \rightarrow N_2(g) + 2H_2O(g)$

    molar masses: N_2H_4, 32.05 g; O_2, 32.00 g; N_2, 28.02 g; H_2O, 18.02 g

$$20.0 \text{ g N}_2\text{H}_4 \times \frac{1 \text{ mol N}_2\text{H}_4}{32.05 \text{ g N}_2\text{H}_4} = 0.624 \text{ mol N}_2\text{H}_4$$

$$20.0 \text{ g O}_2 \times \frac{1 \text{ mol O}_2}{32.00 \text{ g O}_2} = 0.625 \text{ mol O}_2$$

The two reactants are present in nearly the required ratio for complete reaction (due to the 1:1 stoichiometry of the reaction and the very similar molar masses of the substances). We will consider N_2H_4 as the limiting reactant in the following calculations.

$$0.624 \text{ mol N}_2\text{H}_4 \times \frac{1 \text{ mol N}_2}{1 \text{ mol N}_2\text{H}_4} = 0.624 \text{ mol N}_2$$

$$0.624 \text{ mol N}_2 \times \frac{28.02 \text{ g N}_2}{1 \text{ mol N}_2} = 17.5 \text{ g N}_2$$

$$0.624 \text{ mol N}_2\text{H}_4 \times \frac{2 \text{ mol H}_2\text{O}}{1 \text{ mol N}_2\text{H}_4} = 1.248 \text{ mol H}_2\text{O} = 1.25 \text{ mol H}_2\text{O}$$

$$1.248 \text{ mol H}_2\text{O} \times \frac{18.02 \text{ g H}_2\text{O}}{1 \text{ mol H}_2\text{O}} = 22.5 \text{ g H}_2\text{O}$$

92. $$12.5 \text{ g theory} \times \frac{40 \text{ g actual}}{100 \text{ g theory}} = 5.0 \text{ g}$$

Cumulative Review: Chapters 8 and 9

2. On a microscopic basis, one mole of a substance represents Avogadro's number (6.022 × 10²³) of individual units (atoms or molecules) of the substance. On a macroscopic, more practical basis, one mole of a substance represents the amount of substance present when the molar mass of the substance in grams is taken (for example 12.01 g of carbon will be one mole of carbon). Chemists have chosen these definitions so that there will be a simple relationship between measurable amounts of substances (grams) and the actual number of atoms or molecules present, and so that the number of particles present in samples of *different* substances can easily be compared. For example, it is known that carbon and oxygen react by the reaction

$$C(s) + O_2(g) \rightarrow CO_2(g)$$

   Chemists understand this equation to mean that one carbon atom reacts with one oxygen molecule to produce one molecule of carbon dioxide, and also that one mole (12.01 g) of carbon will react with one mole (32.00 g) of oxygen to produce one mole (44.01 g) of carbon dioxide.

4. The molar mass of a compound is the mass in grams of one mole of the compound (6.022 × 10²³ molecules of the compound), and is calculated by summing the average atomic masses of all the atoms present in a molecule of the compound. For example, a molecule of the compound H_3PO_4 contains three hydrogen atoms, one phosphorus atom, and four oxygen atom: the molar mass is obtained by adding up the average atomic masses of these atoms: molar mass H_3PO_4 = 3(1.008 g) + 1(30.97 g) + 4(16.00 g) = 97.99 g

6. The empirical formula of a compound represents the lowest ratio of the relative number of atoms of each type present in a molecule of the compound, whereas the molecular formula represents the actual number of atoms of each type present in a real molecule of the compound. For example, both acetylene (molecular formula C_2H_2) and benzene (molecular formula C_6H_6) have the same relative number of carbon and hydrogen atoms (one hydrogen for each carbon atom), and so have the same empirical formula (CH). Once the empirical formula of a compound has been determined, it is also necessary to determine the molar mass of the compound before the actual molecular formula can be calculated. Since real molecules cannot contain fractional parts of atoms, the molecular formula is always a whole number multiple of the empirical formula. For the examples above, the molecular formula of acetylene is twice the empirical formula, and the molecular formula of benzene is six times the empirical formula (both factors are integers).

8. In question 7, we chose to calculate the percentage composition of phosphoric acid, H_3PO_4: 3.086% H, 31.60% P, 65.31% O. We could convert this percentage composition data into "experimental" data by first choosing a mass of sample to be "analyzed", and then calculating what mass of each element is present in this size sample using the percentage of each element. For example, suppose we choose our sample to have a mass of 2.417 g. Then the masses of H, P, and O present in this sample would be given by the following

$$\text{g H} = (2.417 \text{ g sample}) \times \frac{3.086 \text{ g H}}{100.0 \text{ g sample}} = 0.07459 \text{ g H}$$

$$\text{g P} = (2.417 \text{ g sample}) \times \frac{31.60 \text{ g P}}{100.0 \text{ g sample}} = 0.7638 \text{ g P}$$

$$\text{g O} = (2.417 \text{ g sample}) \times \frac{65.31 \text{ g O}}{100.0 \text{ g sample}} = 1.579 \text{ g O}$$

Note that (0.07459 g + 0.7638 g + 1.579 g) = 2.41739 = 2.417 g.

So our new problem could be worded as follows: "A 2.417-g sample of a compound has been analyzed and was found to contain 0.07459 g H, 0.7638 g of P, and 1.579 g of oxygen. Calculate the empirical formula of the compound".

$$\text{mol H} = (0.07459 \text{ g H}) \times \frac{1 \text{ mol H}}{1.008 \text{ g H}} = 0.07400 \text{ mol H}$$

$$\text{mol P} = (0.7638 \text{ g P}) \times \frac{1 \text{ mol P}}{30.97 \text{ g P}} = 0.02466 \text{ mol P}$$

$$\text{mol O} = (1.579 \text{ g O}) \times \frac{1 \text{ mol O}}{16.00 \text{ g O}} = 0.09869 \text{ mol O}$$

Dividing each of these numbers of moles by the smallest number of moles (0.02466 mol P) gives the following

$$\frac{0.07400 \text{ mol H}}{0.02466} = 3.001 \text{ mol H}$$

$$\frac{0.02466 \text{ mol P}}{0.02466} = 1.000 \text{ mol P}$$

$$\frac{0.09869 \text{ mol O}}{0.02466} = 4.002 \text{ mol O}$$

and the empirical formula is (not surprisingly) just H_3PO_4!

10. The mole ratios for a reaction are based on the *coefficients* of the balanced chemical equation for the reaction: these coefficients show in what proportions molecules (or moles of molecules) combine. For a given amount of C_2H_5OH, the following mole ratios could be constructed, which would enable you to calculate the number of moles of each product, or of the second reactant, that would be involved.

$$C_2H_5OH(l) + 3O_2(g) \rightarrow 2CO_2(g) + 3H_2O(g)$$

to calculate mol CO_2 produced: $\dfrac{2 \text{ mol } CO_2}{1 \text{ mol } C_2H_5OH}$

to calculate mol H_2O produced: $\dfrac{3 \text{ mol } H_2O}{1 \text{ mol } C_2H_5OH}$

to calculate mol O_2 required: $\dfrac{3 \text{ mol } O_2}{1 \text{ mol } C_2H_5OH}$

We could then calculate the numbers of moles of the other substances if 0.65 mol of C_2H_5OH were to be combusted as follows:

mol CO_2 produced = (0.65 mol C_2H_5OH) × $\dfrac{2 \text{ mol } CO_2}{1 \text{ mol } C_2H_5OH}$ = 1.3 mol CO_2

mol H_2O produced = (0.65 mol C_2H_5OH) × $\dfrac{3 \text{ mol } H_2O}{1 \text{ mol } C_2H_5OH}$ = 1.95 = 2.0 mol H_2O

mol O_2 required = (0.65 mol C_2H_5OH × $\dfrac{3 \text{ mol } O_2}{1 \text{ mol } C_2H_5OH}$ = 1.95 = 2.0 mol O_2

12. Although we can calculate specifically the exact amounts of each reactant needed for a chemical reaction, oftentimes reaction mixtures are prepared using more or less arbitrary amounts of the reagents. However, regardless of how much of each reagent may be used for a reaction, the substances still react stoichiometrically, according to the mole ratios derived from the balanced chemical equation for the reaction. When arbitrary amounts of reactants are used, there will be one reactant which, stoichiometrically, is present in the least amount. This substance is called the *limiting reactant* for the experiment. It is the limiting reactant that controls how much product is formed, regardless of how much of the other reactants are present. The limiting reactant limits the amount of product that can form in the experiment, because once the limiting reactant has reacted completely, the reaction must stop. We say that the other reactants in the experiment are present in excess, which means that a portion of these reactants will still be present unchanged after the reaction has ended and the limiting reactant has been used up completely.

14. The *theoretical yield* for an experiment is the mass of product calculated based on the limiting reactant for the experiment being completely consumed. The *actual yield* for an experiment is the mass of product actually collected by the experimenter. Obviously, any experiment is restricted by the skills of the experimenter and by the inherent limitations of the experimental method being used. For these reasons, the actual yield is often *less* than the theoretical yield (most scientific writers report the actual or percentage yield for their experiments as an indication of the usefulness of their experiments). Although one would expect that the actual yield should never be more than the theoretical yield, in real experiments, sometimes this happens: however, an actual yield greater than a theoretical yield is usually taken to mean that something is *wrong* in either the experiment (for example, impurities may

be present, or the reaction may not occur as envisioned) or in the calculations.

16. NH_3: $\%N = \dfrac{14.01 \text{ g N}}{17.03 \text{ g}} \times 100 = 82.27\% \text{ N}$

$\%H = \dfrac{3(1.008 \text{ g H})}{17.03 \text{ g}} \times 100 = 17.76\% \text{ H}$

SiH_4: $\%Si = \dfrac{28.09 \text{ g Si}}{32.12 \text{ g}} \times 100 = 87.45\% \text{ Si}$

$\%H = \dfrac{4(1.008 \text{ g H})}{32.12 \text{ g}} \times 100 = 12.55\% \text{ H}$

K_2CrO_4: $\%K = \dfrac{2(39.10 \text{ g K})}{194.2 \text{ g}} \times 100 = 40.27\% \text{ K}$

$\%Cr = \dfrac{52.00 \text{ g Cr}}{194.2 \text{ g}} \times 100 = 26.78\% \text{ Cr}$

$\%O = \dfrac{4(16.00 \text{ g O})}{194.2 \text{ g}} \times 100 = 32.96\% \text{ O}$

$AuCl_3$: $\%Au = \dfrac{197.0 \text{ g Au}}{303.4 \text{ g}} \times 100 = 64.93\% \text{ Au}$

$\%Cl = \dfrac{3(35.45 \text{ g Cl})}{303.4 \text{ g}} \times 100 = 35.05\% \text{ Cl}$

$KMnO_4$: $\%K = \dfrac{39.10 \text{ g K}}{158.0 \text{ g}} \times 100 = 24.75\% \text{ K}$

$\%Mn = \dfrac{54.94 \text{ g Mn}}{158.0 \text{ g}} \times 100 = 34.77\% \text{ Mn}$

$\%O = \dfrac{4(16.00 \text{ g O})}{158.0 \text{ g}} \times 100 = 40.51\% \text{ O}$

$KClO_3$: $\%K = \dfrac{39.10 \text{ g K}}{122.5 \text{ g}} \times 100 = 31.92\% \text{ K}$

$\%Cl = \dfrac{35.45 \text{ g Cl}}{122.5 \text{ g}} \times 100 = 28.94\% \text{ Cl}$

$\%O = \dfrac{3(16.00 \text{ g O})}{122.5 \text{ g}} \times 100 = 39.18\% \text{ O}$

18. a. $2AgNO_3(aq) + CaSO_4(aq) \rightarrow Ag_2SO_4(s) + Ca(NO_3)_2(aq)$

Molar masses: $AgNO_3$, 169.9 g; Ag_2SO_4, 311.9 g; $Ca(NO_3)_2$, 164.1 g

$$25.0 \text{ g AgNO}_3 \times \frac{1 \text{ mol}}{169.9 \text{ g}} = 0.147 \text{ mol AgNO}_3$$

$$0.147 \text{ mol AgNO}_3 \times \frac{1 \text{ mol Ag}_2\text{SO}_4}{2 \text{ mol AgNO}_3} = 0.0735 \text{ mol Ag}_2\text{SO}_4$$

$$0.0735 \text{ mol Ag}_2\text{SO}_4 \times \frac{311.9 \text{ g}}{1 \text{ mol}} = 22.9 \text{ g Ag}_2\text{SO}_4$$

$$0.147 \text{ mol AgNO}_3 \times \frac{1 \text{ mol Ca(NO}_3)_2}{2 \text{ mol AgNO}_3} = 0.0735 \text{ mol Ca(NO}_3)_2$$

$$0.0735 \text{ mol Ca(NO}_3)_2 \times \frac{164.1 \text{ g}}{1 \text{ mol}} = 12.1 \text{ g Ca(NO}_3)_2$$

b. $2\text{Al}(s) + 6\text{HNO}_3(aq) \rightarrow 2\text{Al(NO}_3)_3(aq) + 3\text{H}_2(g)$

Molar masses: Al, 26.98 g; Al(NO$_3$)$_3$, 213.0 g; H$_2$, 2.016 g

$$25.0 \text{ g Al} \times \frac{1 \text{ mol}}{26.98 \text{ g}} = 0.927 \text{ mol Al}$$

$$0.927 \text{ mol Al} \times \frac{2 \text{ mol Al(NO}_3)_3}{2 \text{ mol Al}} = 0.927 \text{ mol Al(NO}_3)_3$$

$$0.927 \text{ mol Al(NO}_3)_3 \times \frac{213.0 \text{ g}}{1 \text{ mol}} = 197 \text{ g Al(NO}_3)_3$$

$$0.927 \text{ mol Al} \times \frac{3 \text{ mol H}_2}{2 \text{ mol Al}} = 1.39 \text{ mol H}_2$$

$$1.39 \text{ mol H}_2 \times \frac{2.016 \text{ g}}{1 \text{ mol}} = 2.80 \text{ g H}_2$$

c. $\text{H}_3\text{PO}_4(aq) + 3\text{NaOH}(aq) \rightarrow \text{Na}_3\text{PO}_4(aq) + 3\text{H}_2\text{O}(l)$

Molar masses: H$_3$PO$_4$, 97.99 g; Na$_3$PO$_4$, 163.9 g; H$_2$O, 18.02 g

$$25.0 \text{ g H}_3\text{PO}_4 \times \frac{1 \text{ mol}}{97.99 \text{ g}} = 0.255 \text{ mol H}_3\text{PO}_4$$

$$0.255 \text{ mol H}_3\text{PO}_4 \times \frac{1 \text{ mol Na}_3\text{PO}_4}{1 \text{ mol H}_3\text{PO}_4} = 0.255 \text{ mol Na}_3\text{PO}_4$$

$$0.255 \text{ mol Na}_3\text{PO}_4 \times \frac{163.9 \text{ g}}{1 \text{ mol}} = 41.8 \text{ g Na}_3\text{PO}_4$$

$$0.255 \text{ mol H}_3\text{PO}_4 \times \frac{3 \text{ mol H}_2\text{O}}{1 \text{ mol H}_3\text{PO}_4} = 0.765 \text{ mol H}_2\text{O}$$

$$0.765 \text{ mol } H_2O \times \frac{18.02 \text{ g}}{1 \text{ mol}} = 13.8 \text{ g } H_2O$$

d. $CaO(s) + 2HCl(aq) \rightarrow CaCl_2(aq) + H_2O(l)$

Molar masses: CaO, 56.08 g; $CaCl_2$, 111.0 g; H_2O, 18.02 g

$$25.0 \text{ g CaO} \times \frac{1 \text{ mol}}{56.08 \text{ g}} = 0.446 \text{ mol CaO}$$

$$0.446 \text{ mol CaO} \times \frac{1 \text{ mol } CaCl_2}{1 \text{ mol CaO}} = 0.446 \text{ mol } CaCl_2$$

$$0.446 \text{ mol } CaCl_2 \times \frac{111.0 \text{ g}}{1 \text{ mol}} = 49.5 \text{ g } CaCl_2$$

$$0.446 \text{ mol CaO} \times \frac{1 \text{ mol } H_2O}{1 \text{ mol CaO}} = 0.446 \text{ mol } H_2O$$

$$0.446 \text{ mol } H_2O \times \frac{18.02 \text{ g}}{1 \text{ mol}} = 8.04 \text{ g } H_2O$$

20.   For potassium: $2K(s) + Cl_2(g) \rightarrow 2KCl(s)$

Molar masses: K, 39.10 g; Cl_2, 70.90 g; KCl, 74.55 g

$$25.0 \text{ g K} \times \frac{1 \text{ mol}}{39.10 \text{ g}} = 0.639 \text{ mol K}$$

$$50.0 \text{ g } Cl_2 \times \frac{1 \text{ mol}}{70.90 \text{ g}} = 0.705 \text{ mol } Cl_2$$

K is the limiting reactant

$$0.639 \text{ mol K} \times \frac{2 \text{ mol KCl}}{2 \text{ mol K}} \times \frac{74.55 \text{ g}}{1 \text{ mol}} = 47.6 \text{ g KCl}$$

For calcium: $Ca(s) + Cl_2(g) \rightarrow CaCl_2(s)$

Molar masses: Ca, 40.08 g; Cl_2, 70.90 g; $CaCl_2$, 111.0 g

$$25.0 \text{ g Ca} \times \frac{1 \text{ mol}}{40.08 \text{ g}} = 0.624 \text{ mol Ca}$$

$$50.0 \text{ g } Cl_2 \times \frac{1 \text{ mol}}{70.90 \text{ g}} = 0.705 \text{ mol } Cl_2$$

Ca is the limiting reactant

$$0.624 \text{ mol Ca} \times \frac{1 \text{ mol } CaCl_2}{1 \text{ mol Ca}} \times \frac{111.0 \text{ g}}{1 \text{ mol}} = 69.3 \text{ g } CaCl_2$$

For aluminum: $2Al(s) + 3Cl_2(g) \rightarrow 2AlCl_3(s)$

Molar masses: Al, 26.98 g; Cl_2 70.90 g; $AlCl_3$, 133.3 g

$$25.0 \text{ g Al} \times \frac{1 \text{ mol}}{26.98 \text{ g}} = 0.927 \text{ mol Al}$$

$$50.0 \text{ g Cl}_2 \times \frac{1 \text{ mol}}{70.90 \text{ g}} = 0.705 \text{ mol Cl}_2$$

Cl_2 is the limiting reactant

$$0.705 \text{ mol Cl}_2 \times \frac{2 \text{ mol AlCl}_3}{3 \text{ mol Cl}_2} \times \frac{133.3 \text{ g}}{1 \text{ mol}} = 62.7 \text{ g AlCl}_3$$

Chapter 10   Modern Atomic Theory

2. The different forms of electromagnetic radiation are similar in that they all exhibit the same type of wave-like behavior and are propagated through space at the same speed (the speed of light). The types of electromagnetic radiation differ in their frequency (and wavelength) and in the resulting amount of energy carried per photon.

4. The *speed* of electromagnetic radiation represents how fast a given wave moves through space. The *frequency* of electromagnetic radiation represents how many complete cycles of the wave pass a given point per second. For example, your favorite radio station broadcasts waves of a particular frequency that distinguishes it, but how long those waves take to reach you depends on their speed through space.

6. Although electromagnetic radiation exhibits characteristic wave-like properties, such radiation also demonstrates a particle-like nature. For example, when an excited atom emits radiation, the radiation demonstrates wave properties, but occurs in discrete particle-like bundles (photons). The wave-particle nature of light refers to the fact that a beam of electromagnetic energy can be considered not only as a continuous wave, but also as a stream of discrete packets of energy moving through space.

8. A photon having an energy corresponding to the energy difference between the two states is emitted by an atom in an excited state when it returns to its ground state.

10. smaller ($E = h\nu$)

12. When excited hydrogen atoms emit their excess energy, the photons of radiation emitted are always of exactly the same wavelength and energy. We consider this to mean that the hydrogen atom possesses only certain allowed energy states, and that the photons emitted correspond to the atom changing from one of these allowed energy states to another of the allowed energy state. The energy of the photon emitted corresponds to the energy difference in the allowed states. If the hydrogen atom did not possess discrete energy levels, then we would expect the photons emitted to have random wavelengths and energies.

14. difference

16. The ground state of an atom is its lowest possible energy state.

18. According to Bohr, electrons move in discrete, fixed circular *orbits* around the nucleus. If the wavelength of the applied energy corresponds to the *difference in energy* between the two orbits, the atom absorbs a photon and the electron moves to a larger orbit.

20. Bohr's theory *explained* the experimentally *observed* line spectrum of hydrogen *exactly*. Bohr's theory was ultimately discarded because when attempts were made to extend the theory to atoms other than hydrogen,

the calculated properties did *not* correspond closely to experimental measurements.

22. An orbit represents a definite, exact circular pathway around the nucleus in which an electron can be found. An orbital represents a region of space in which there is a high probability of finding the electron.

24. Any experiment which sought to measure the exact location of an electron (such as shooting a beam of light at it) would cause the electron to move. Any measurement made would necessitate the application or removal of energy, which would disturb the electron from where it had been before the measurement.

26. Pictures we draw to represent orbitals should only be interpreted as probability maps. They are not meant to represent that the electron moves only on the surface of, or within, the region drawn in the picture. Since the mathematical probability of finding the electron never actually becomes zero on moving outward from the nucleus, scientists have decided that pictures of orbitals should represent a 90% probability that the electron will be found inside the region depicted in the drawing (for 100% probability, the orbital would have to encompass all space).

28. The *p* orbitals, in general, have two lobes and are sometimes described as having a "dumbbell" shape. The 2*p* and 3*p* orbitals are similar in shape, and in the fact that there are three equivalent 2*p* or 3*p* orbitals in the 2*p* or 3*p* subshell. The orbitals differ in size, mean distance from the nucleus, and energy.

30. increases; the energy of an electron increases with its principal quantum number.

32. The fourth principal energy level is divided into four sublevels: these sublevels are given the designations 4*s*, 4*p*, 4*d*, and 4*f*. The fifth principal energy level is divided into five sublevers: these are given the designations 5*s*, 5*p*, 5*d*, 5*f*, and 5*g*.

34. The Pauli exclusion principle states that an orbital can hold a maximum of two electrons, and those two electrons must have opposite spins.

36. increases; as you move out from the nucleus, there is more space and room for more sublevels.

38. opposite

40. a. incorrect (the *n* = 1 shell has only the 1*s* subshell)
    b. incorrect (the *n* = 2 shell has only 2*s* and 2*p* subshells)
    c. correct (the *n* = 4 shell contains *s*, *p*, *d* and *f* subshells)
    d. correct (the *n* = 5 shell contains *s*, *p*, *d*, *f*, and *g* subshells)

42. When a hydrogen atom is in its ground state, the electron is found in the 1s orbital. The 1s orbital has the lowest energy of all the possible hydrogen orbitals.

44. The elements in a given vertical column of the periodic table have the same valence electron configuration. Having the same valence electron configuration causes the elements in a given group to have similar chemical properties.

46. a. $1s^2\ 2s^2\ 2p^6\ 3s^2\ 3p^6\ 4s^2\ 3d^{10}\ 4p^6\ 5s^2$
    b. $1s^2\ 2s^2\ 2p^6\ 3s^2\ 3p^6\ 4s^2\ 3d^{10}$
    c. $1s^2$
    d. $1s^2\ 2s^2\ 2p^6\ 3s^2\ 3p^6\ 4s^2\ 3d^{10}\ 4p^5$

48. a. $1s^2\ 2s^2\ 2p^6\ 3s^2\ 3p^6\ 4s^2$
    b. $1s^2\ 2s^2\ 2p^6\ 3s^2\ 3p^6\ 4s^1$
    c. $1s^2\ 2s^2\ 2p^5$
    d. $1s^2\ 2s^2\ 2p^6\ 3s^2\ 3p^6\ 4s^2\ 3d^{10}\ 4p^6$

50. a. (↑↓) (↑↓) (↑↓)(↑↓)(↑↓) (↑↓) (↑)( )( )
         1s   2s   2p   3s   3p

    b. (↑↓) (↑↓) (↑↓)(↑↓)(↑↓) (↑↓) (↑)(↑)(↑)
         1s   2s   2p   3s   3p

    c. (↑↓) (↑↓) (↑↓)(↑↓)(↑↓) (↑↓) (↑↓)(↑↓)(↑↓) (↑↓)
         1s   2s   2p   3s   3p   4s

        (↑↓)(↑↓)(↑↓)(↑↓)(↑↓) (↑↓)(↑↓)(↑)
              3d   4p

    d. (↑↓) (↑↓) (↑↓)(↑↓)(↑↓) (↑↓) (↑↓)(↑↓)(↑↓)
         1s   2s   2p   3s   3p

52. For the representative elements (those filling s and p subshells), the group number gives the number of valence electrons.

    a. one (3s)
    b. two (4s)
    c. seven (5s, 5p)
    d. five (2s, 2p)

54. The properties of Rb and Sr suggest that they are members of Groups 1 and 2, respectively, and so must be filling the 5s orbital. The 5s orbital is lower in energy (and fills before) the 4d orbitals.

56. a. [Ar] $4s^2$
    b. [Rn] $7s^1$
    c. [Kr] $5s^2\ 4d^1$
    d. [Xe] $6s^2\ 4f^1\ 5d^1$

58. a. [Ne] $3s^2\ 3p^3$
    b. [Ne] $3s^2\ 3p^5$
    c. [Ne] $3s^2$
    d. [Ar] $4s^2\ 3d^{10}$

60. a. one
    b. two
    c. zero
    d. ten

62. Figure 10.27 shows the orbitals being filled as a function of location in the periodic table.

    a. 5f
    b. 5f
    c. 4f
    d. 6p

64. Figure 10.30 shows partial electronic configurations.
    a. [Rn] $7s^2\ 5f^3\ 6d^1$
    b. [Ar] $4s^2\ 3d^5$
    c. [Xe] $6s^2\ 4f^{14}\ d^{10}$
    d. [Rn] $7s^1$

66. The metallic elements *lose* electrons and form *positive* ions (cations); the nonmetallic elements *gain* electrons and form *negative* ions (anions). Remember that the electron itself is *negatively* charged.

68. All exist as *diatomic* molecules (F_2, Cl_2, Br_2, I_2); all are *non*metals; all have relatively high electronegativities; all form 1- ions in reacting with metallic elements.

70. Elements at the *left* of a period (horizontal row) lose electrons more readily; at the left of a period (given principal energy level) the nuclear charge is the smallest and the electrons are least tightly held.

72. The elements of a given period (horizontal row) have valence electrons in the same principal energy level. Nuclear charge, however, increases across a period going from left to right. Atoms at the left side have smaller nuclear charges, and hold onto their valence electrons less tightly.

74. The *nuclear charge* increases from left to right within a period, pulling progressively more tightly on the valence electrons.

76. Ionization energies decrease in going from top to bottom within a vertical group; ionization energies increase in going from left to right within a horizontal period.

   a. Li
   b. Ca
   c. Cl
   d. S

78. Atomic size increases in going from top to bottom within a vertical group; atomic size decreases in going from left to right within a horizontal period.

   a. Na
   b. S
   c. N
   d. F

80. speed of light

82. photons

84. quantized

86. orbital

88. transition metal

90. spins

92. a. $1s^2\ 2s^2\ 2p^6\ 3s^2\ 3p^6\ 4s^1$ [Ar] $4s^1$

   (↑↓) (↑↓) (↑↓)(↑↓)(↑↓) (↑↓) (↑↓)(↑↓)(↑↓) (↑)
   1s   2s   2p   3s   3p   4s

   b. $1s^2\ 2s^2\ 2p^6\ 3s^2\ 3p^6\ 4s^2\ 3d^2$ [Ar] $4s^2\ 3d^2$

   (↑↓) (↑↓) (↑↓)(↑↓)(↑↓) (↑↓) (↑↓)(↑↓)(↑↓) (↑↓) (↑)(↑)( )( )( )
   1s   2s   2p   3s   3p   4s   3d

   c. $1s^2\ 2s^2\ 2p^6\ 3s^2\ 3p^2$ [Ne] $3s^2\ 3p^2$

   (↑↓) (↑↓) (↑↓)(↑↓)(↑↓) (↑↓) (↑)(↑)( )
   1s   2s   2p   3s   3p

   d. $1s^2\ 2s^2\ 2p^6\ 3s^2\ 3p^6\ 4s^2\ 3d^6$ [Ar] $4s^2\ 3d^6$

   (↑↓) (↑↓) (↑↓)(↑↓)(↑↓) (↑↓) (↑↓)(↑↓)(↑↓) (↑↓) (↑↓)(↑)(↑)(↑)(↑)
   1s   2s   2p   3s   3p   4s   3d

   e. $1s^2\ 2s^2\ 2p^6\ 3s^2\ 3p^6\ 4s^2\ 3d^{10}$ [Ar] $4s^2\ 3d^{10}$

   (↑↓) (↑↓) (↑↓)(↑↓)(↑↓) (↑↓) (↑↓)(↑↓)(↑↓) (↑↓) (↑↓)(↑↓)(↑↓)(↑↓)(↑↓)
   1s   2s   2p   3s   3p   4s   3d

94. a. ns^2 b. $ns^2\ np^5$
    c. $ns^2\ np^4$ d. ns^1
    e. $ns^2\ np^4$

96. a. $$\lambda = \frac{h}{mv}$$

   $$\lambda = \frac{6.63 \times 10^{-34}\ J\ s}{(9.11 \times 10^{-31}\ kg)[0.90(3.00 \times 10^8\ m\ s^{-1})]}$$

   $\lambda = 2.7 \times 10^{-12}$ m (0.0027 nm)

   b. 4.4×10^{-34} m

   c. 2×10^{-35} m

   The wavelengths for the ball and the person are *infinitesimally small*, whereas the wavelength for the electron is nearly the same order of magnitude as the diameter of a typical atom.

98. Light is emitted from the hydrogen atom only at certain fixed wavelengths. If the energy levels of hydrogen were *continuous*, a hydrogen atom would emit energy at all possible wavelengths.

132 Chapter 10 Modern Atomic Theory

100. The third principal energy level of hydrogen is divided into *three* sublevels (3s, 3p, and 3d); there is a *single* 3s orbital; there is a set of *three* 3p orbitals; there is a set of *five* 3d orbitals. See Figures 10.21-10.24 for the shapes of these orbitals.

102. a. incorrect; the $n = 1$ energy shell has only the 1s subshell
      b. correct
      c. incorrect; the $n = 3$ energy shell has only 3s. 3p, and 3d subshells
      d. correct
      e. correct
      f. correct

104. a. $1s^2\ 2s^2\ 2p^6\ 3s^2\ 3p^6\ 4s^2\ 3d^{10}\ 4p^5$
      b. $1s^2\ 2s^2\ 2p^6\ 3s^2\ 3p^6\ 4s^2\ 3d^{10}\ 4p^6\ 5s^2\ 4d^{10}\ 5p^6$
      c. $1s^2\ 2s^2\ 2p^6\ 3s^2\ 3p^6\ 4s^2\ 3d^{10}\ 4p^6\ 5s^2\ 4d^{10}\ 5p^6\ 6s^2$
      d. $1s^2\ 2s^2\ 2p^6\ 3s^2\ 3p^6\ 4s^2\ 3d^{10}\ 4p^4$

106. a. five (2s, 2p)
      b. seven (3s, 3p)
      c. one (3s)
      d. three (3s, 3p)

108. a. [Kr] $5s^2\ 4d^2$ b. [Kr] $5s^2\ 4d^{10}\ 5p^5$
      c. [Ar] $4s^2\ 3d^{10}\ 4p^2$ d. [Xe] $6s^1$

110. The *position* of the element (both in terms of the vertical column and the horizontal row) tells you which set of orbitals is being filled last. See Figure 10.27 for details.

      a. 3d   b. 4d
      c. 5f   d. 4p

112. metals, low; nonmetals, high

114. Atomic size increases in going from top to bottom within a vertical group; atomic size decreases in going from left to right within a horizontal period.

      a. Ca
      b. P
      c. K

Chapter 11   Chemical Bonding

2. The strength of a chemical bond is characterized by the bond energy (the amount of energy required to break the bond).

4. *Ionic* bonding results from the complete transfer of an electron from one atom to another, whereas *covalent* bonding exists when two atoms share pairs of electrons.

6. The HF molecule contains atoms of two different elements. In bonding with each other, these atoms share a pair of valence electrons, but they do not share them equally. The bonding in HF is described as polar covalent, which indicates that although valence electron pairs are shared between atoms, the electron pair is attracted more strongly by one of the atoms than the other. Nonetheless, an HF molecule is more stable in energy than are separated H and F atoms.

8. A polar covalent bond results when one atom of the bond attracts electrons more strongly toward itself than does the second atom of the bond. A polar covalent bond results when the atoms forming the covalent bond have different electronegativities. The fact that the bonds in a molecule are polar does *not* necessarily mean that the overall molecule itself will be polar: the overall polarity of the molecule also depends on the geometry of the molecule.

10. electronegativity

12. In each case, the element higher up within a given group of the periodic table, or to the right within a period, has the higher electronegativity.

    a.   K < Ca < Sc

    b.   At < Br < F

    c.   C < N < O

14. Generally, covalent bonds between atoms of *different* elements are *polar*.

    a.   covalent

    b.   polar covalent

    c.   polar covalent

    d.   ionic

16. For a bond to be polar covalent, the atoms involved in the bond must have different electronegativities (must be of different elements).

    a.   covalent (atoms of the same element)

    b.   covalent (atoms of the same element)

    c.   polar covalent (different elements)

    d.   polar covalent (different elements)

18. The *degree* of polarity of a polar covalent bond is indicated by the magnitude of the difference in electronegativities of the elements involved: the larger the difference in electronegativity, the more polar is the bond. Electronegativity differences are given in parentheses below:

   a.   H-O (1.4); H-N (0.9); the H-O bond is more polar.

   b.   H-N (0.9); H-F (1.9); the H-F bond is more polar.

   c.   H-O (1.4); H-F (1.9); the H-F bond is more polar.

   d.   H-O (1.4); H-Cl (0.9); the H-O bond is more polar.

20. The greater the electronegativity difference between two atoms, the more ionic will be the bond between those two atoms. Electronegativity differences are given in parentheses.

   a.   Na-O (2.6) has more ionic character than Na-N (2.1)

   b.   K-S (1.7) has more ionic character than K-P (1.3)

   c.   K-Cl (2.2) has more ionic character than Na-Cl (2.1)

   d.   Na-Cl (2.1) has more ionic character than Mg-Cl (1.8)

22. The presence of strong bond dipoles and a large overall dipole moment in water make it a very polar substance overall. Among those properties of water that are dependent on its dipole moment are its freezing point, melting point, vapor pressure, and its ability to dissolve many substances.

24. In a diatomic molecule containing two different elements, the more electronegative atom will be the negative end of the molecule, and the *less* electronegative atom will be the positive end.

   a.   H

   b.   Cl

   c.   I

26. In the figures, the arrow points toward the more electronegative atom.

   a. $\delta+$ P→F $\delta-$

   b. $\delta+$ P→O $\delta-$

   c. $\delta+$ P→C $\delta-$

   d.   P and H have similar electronegativities.

28. In the figures, the arrow points toward the more electronegative atom.

    a. δ+ P→S δ−

    b. δ+ S→O δ−

    c. δ+ S→N δ−

    d. δ+ S→Cl δ−

30. previous

32. Atoms in covalent molecules gain a configuration like that of a noble gas by sharing one or more pairs of electrons between atoms: such shared pairs of electrons "belong" to each of the atoms of the bond at the same time. In ionic bonding, one atom completely gives over one or more electrons to another atom, and then the resulting ions behave independently of one another.

34. a. Li $1s^2\ 2s^1$

    Li$^+$ $1s^2$

    He has the same configuration as Li$^+$

    b. Br $1s^2\ 2s^2\ 2p^6\ 3s^2\ 3p^6\ 4s^2\ 3d^{10}\ 4p^5$

    Br$^-$ $1s^2\ 2s^2\ 2p^6\ 3s^2\ 3p^6\ 4s^2\ 3d^{10}\ 4p^6$

    Kr has the same configuration as Br$^-$

    c. Cs $1s^2\ 2s^2\ 2p^6\ 3s^2\ 3p^6\ 4s^2\ 3d^{10}\ 4p^6\ 5s^2\ 4d^{10}\ 5p^6\ 6s^1$

    Cs$^+$ $1s^2\ 2s^2\ 2p^6\ 3s^2\ 3p^6\ 4s^2\ 3d^{10}\ 4p^6\ 5s^2\ 4d^{10}\ 5p^6$

    Xe has the same configuration as Cs$^+$

    d. S $1s^2\ 2s^2\ 2p^6\ 3s^2\ 3p^4$

    S^{2-} $1s^2\ 2s^2\ 2p^6\ 3s^2\ 3p^6$

    Ar has the same configuration as S^{2-}

    e. Mg $1s^2\ 2s^2\ 2p^6\ 3s^2$

    Mg^{2+} $1s^2\ 2s^2\ 2p^6$

    Ne has the same configuration as Mg^{2+}

36. a. Ca^{2+} (Ca has two electrons more than the noble gas Ar)

    b. N^{3-} (N has three electrons fewer than the noble gas Ne)

c. Br⁻ (Br has one electron fewer than the noble gas Kr)

d. Mg^{2+} (Mg has two electrons more than the noble gas Ne)

38. a. Na_2S — Na has one electrone more than a noble gas; S has two electrons fewer than a noble gas.

b. BaSe — Ba has two electrons more than a noble gas; Se has two electrons fewer than a noble gas.

c. $MgBr_2$ — Mg has two electrons more than a noble gas; Br has one electron less than a noble gas.

d. Li_3N — Li has one electron more than a noble gas; N has three electrons fewer than a noble gas.

e. KH — K has one electron more than a noble gas; H has one electron less than a noble gas.

40. a. Al^{3+}, [Ne]; S^{2-}, [Ar]
    b. Mg^{2+}, [Ne]; N^{3-}, [Ne]
    c. Rb^{+}, [Kr]; O^{2-}, [Ne]
    d. Cs^{+}, [Xe]; I^{-}, [Xe]

42. An ionic solid such as NaCl basically consists of an array of alternating positively and negatively charged ions: that is, each positive ion has as its nearest neighbors a group of negative ions, and each negative ion has a group of positive ions surrounding it. In most ionic solids, the ions are packed as tightly as possible.

44. In forming an anion, an atom gains additional electrons in its outermost (valence) shell. Having additional electrons in the valence shell increases the repulsive forces between electrons, and the outermost shell becomes larger to accommodate this.

46. Relative ionic sizes are indicated in Figure 11.9. Within a given horizontal row of the periodic chart, negative ions tend to be larger than positive ions because the negative ions contain a larger number of electrons in the valence shell. Within a vertical group of the periodic table, ionic size increases from top to bottom. In general, positive ions are

Chapter 11   Chemical Bonding   137

smaller than the atoms they come from, whereas negative ions are larger than the atoms they come from.

   a. F^-

   b. Cl^-

   c.   Ca

   d. I^-

48. Relative ionic sizes are indicated in Figure 11.9. Within a given horizontal row of the periodic chart, negative ions tend to be larger than positive ions because the negative ions contain a larger number of electrons in the valence shell. Within a vertical group of the periodic table, ionic size increases from top to bottom. In general, positive ions are smaller than the atoms they come from, whereas negative ions are larger than the atoms they come from.

   a. I^-

   b. Cl^-

   c. Cl^-

   d. S^{2-}

50. When atoms form covalent bonds, they try to attain a valence electronic configuration similar to that of the following noble gas element. When the elements in the first few horizontal rows of the periodic table form covalent bonds, they will attempt to gain configurations similar to the noble gases helium (2 valence electrons, duet rule), and neon and argon (8 valence electrons, octet rule).

52. These elements attain a total of eight valence electrons, making the valence electron configurations similar to those of the noble gases Ne and Ar.

54. When two atoms in a molecule are connected by a triple bond, we are indicating that the atoms share three pairs of electrons (6 electrons) in completing their outermost shells. A simple molecule containing a triple bond is acetylene, C_2H_2 H:C:::C:H

56.   a.   Rb·

      b.   :C̈l·

      c.   :K̈r:

      d.   Ba:

138 Chapter 11   Chemical Bonding

    e. 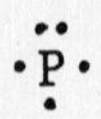   ·P̈·
           ·

    f.    :Ät·

58. a. C provides 4; each Br provides 7; total valence electrons = 32

    b. N provides 5; each O provides 6; total valence electrons = 17

    c. each C provides 4; each H provides 1; total valence electrons = 30

    d. each O provides 6; each H provides 1; total valence electrons = 14

60. a. NH_3 N provides 5 valence electrons.
                 Each H provides 1 valence electron.
                 Total valence electrons = 8

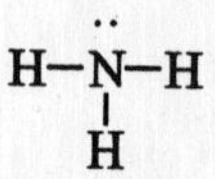

$$H-\overset{..}{N}-H$$
$$|$$
$$H$$

    b. CI_4 C provides 4 valence electrons.
                 Each I provides 7 valence electrons.
                 Total valence electrons = 32

:Ï:
|
:Ï — C — Ï:
|
:Ï:

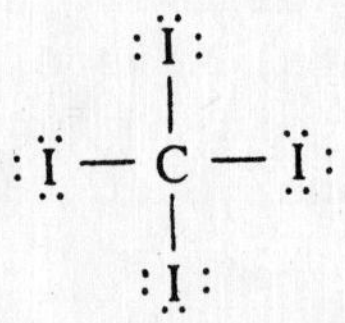

    c. NCl_3 N provides 5 valence electrons.
                 Each Cl provides 7 valence electrons.
                 Total valence electrons = 26.

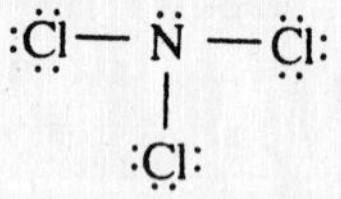

:C̈l — Ṅ — C̈l:
|
:C̈l:

    d. $SiBr_4$ Si provides 4 valence electrons.
                 Each Br provides 7 valence electrons.
                 Total valence electrons = 32

:B̈r:
|
:B̈r — Si — B̈r:
|
:B̈r:

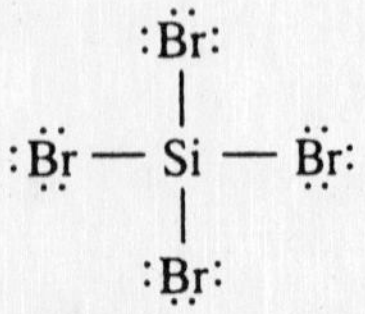

Chapter 11   Chemical Bonding   139

62. a. H_2S Each H provides 1 valence electron.
S provides 6 valence electrons.
Total valence electrons = 8

$$H-\ddot{\underset{..}{S}}-H$$

b. SiF_4 Si provides 4 valence electrons.
Each F provides 7 valence electrons.
Total valence electrons = 32

$$:\ddot{\underset{..}{F}}-\underset{\underset{:\ddot{F}:}{|}}{\overset{\overset{:\ddot{F}:}{|}}{Si}}-\ddot{\underset{..}{F}}:$$

c. C_2H_4 Each C provides 4 valence electrons.
Each H provides 1 valence electron.
Total valence electrons = 12

$$\begin{array}{c} H \quad H \\ | \quad | \\ C=C \\ | \quad | \\ H \quad H \end{array}$$

d. C_3H_8 Each C provides 4 valence electrons.
Each H provides 1 valence electron.
Total valence electrons = 20

$$\begin{array}{c} H \quad H \quad H \\ | \quad | \quad | \\ H-C-C-C-H \\ | \quad | \quad | \\ H \quad H \quad H \end{array}$$

64. a. NO_2 N provides 5 valence electrons.
Each O provides 6 valence electrons.
Total valence electrons = 17

$$\ddot{\underset{..}{O}}=\dot{N}-\ddot{\underset{..}{O}}: \qquad \ddot{\underset{..}{O}}=N-\dot{\underset{..}{O}}:$$

$$\cdot\ddot{\underset{..}{O}}-N=\ddot{\underset{..}{O}} \qquad :\ddot{\underset{..}{O}}-\dot{N}=\ddot{\underset{..}{O}}$$

Note that there are several Lewis structures possible because of the odd number of electrons (the unpaired electron can be located on different atoms in the Lewis structure).

140   Chapter 11   Chemical Bonding

b. H_2SO_4 Each H provides 1 valence electron.
S provides 6 valence electrons.
Each O provides 6 valence electrons.
Total valence electrons = 32

$$H-\ddot{\underset{..}{O}}-\underset{\underset{:\ddot{O}:}{\overset{:\ddot{O}:}{|}}}{S}-\ddot{\underset{..}{O}}-H$$

c. N_2O_4 Each N provides 5 valence electrons.
Each O provides 6 valence electrons.
Total valence electrons = 34

$$:\ddot{O}=N-\underset{\underset{:\ddot{O}:}{|}}{\overset{\overset{:\ddot{O}:}{|}}{N}}=\ddot{O}: \qquad :\ddot{O}-\underset{\underset{:\ddot{O}:}{||}}{\overset{\overset{:O:}{||}}{N}}-N-\ddot{O}:$$

66. a. ClO_3^- Cl provides 7 valence electrons.
Each O provides 6 valence electrons.
The 1- charge means 1 additional electron.
Total valence electrons = 26

$$\left[:\ddot{\underset{..}{Cl}}-\underset{\underset{:\ddot{Cl}:}{|}}{\ddot{O}}-\ddot{\underset{..}{Cl}}:\right]^{1-}$$

b. O_2^{2-} Each O provides 6 valence electrons.
The 2- charge means two additional valence electrons.
Total valence electrons = 14

$$\left[:\ddot{\underset{..}{O}}-\ddot{\underset{..}{O}}:\right]^{2-}$$

c. $C_2H_3O_2^-$ Each C provides 4 valence electrons.
Each H provides 1 valence electron.
Each O provides 6 valence electrons.
The 1- charge means 1 additional valence electron.
Total valence electrons = 24

$$\left[\underset{\underset{H}{|}}{\overset{\overset{H}{|}}{H-C}}-\underset{}{\overset{\overset{:O:}{||}}{C}}-\ddot{O}:\right]^{1-} \quad \left[\underset{\underset{H}{|}}{\overset{\overset{H}{|}}{H-C}}-\underset{}{\overset{\overset{:\ddot{O}:}{|}}{C}}=\ddot{O}\right]^{1-}$$

68. a. HPO_4^{2-}

H provides 1 valence electron.
P provides 5 valence electrons.
Each O provides 6 valence electrons
The 2- charge means two additional valence electrons.
Total valence electrons = 32

$$\left[\begin{array}{c} \ddot{\underset{\cdot\cdot}{\text{O}}}\text{:} \\ | \\ H-\ddot{\underset{\cdot\cdot}{\text{O}}}-P-\ddot{\underset{\cdot\cdot}{\text{O}}}\text{:} \\ | \\ \ddot{\underset{\cdot\cdot}{\text{O}}}\text{:} \end{array} \right]^{2-}$$

b. $H_2PO_4^-$

Each H provides 1 valence electron.
P provides 5 valence electrons.
Each O provides 6 valence electrons.
The 1- charge means one additional valence electron.
Total valence electrons = 32

$$\left[\begin{array}{c} \ddot{\underset{\cdot\cdot}{\text{O}}}\text{:} \\ | \\ H-\ddot{\underset{\cdot\cdot}{\text{O}}}-P-\ddot{\underset{\cdot\cdot}{\text{O}}}-H \\ | \\ \ddot{\underset{\cdot\cdot}{\text{O}}}\text{:} \end{array} \right]^{1-}$$

c. PO_4^{3-}

P provides 5 valence electrons.
Each O provides 6 valence electrons.
The 3- charge means three additional valence electrons.
Total valence electrons = 32

$$\left[\begin{array}{c} \phantom{:\ddot{\text{O}}-}\ddot{\underset{\cdot\cdot}{\text{O}}}\text{:} \\ \phantom{:\ddot{\text{O}}-}| \\ \text{:}\ddot{\underset{\cdot\cdot}{\text{O}}}-P-\ddot{\underset{\cdot\cdot}{\text{O}}}\text{:} \\ \phantom{:\ddot{\text{O}}-}| \\ \phantom{:\ddot{\text{O}}-}\ddot{\underset{\cdot\cdot}{\text{O}}}\text{:} \end{array} \right]^{3-}$$

70. The geometric structure of NH_3 is that of a trigonal pyramid. The nitrogen atom of NH_3 is surrounded by four electron pairs (three are bonding, one is a lone pair). The H-N-H bond angle is somewhat less than 109.5° (due to the presence of the lone pair).

72. The geometric structure of CH_4 is that of a tetrahedron. The carbon atom of CH_4 is surrounded by four bonding electron pairs. The H-C-H bond angle is the characteristic angle of the tetrahedron, 109.5°.

74. The general molecular structure of a molecule is determined by *how many electron pairs* surround the central atom in the molecule, and by which of those electron pairs are used for *bonding* to the other atoms of the molecule. Nonbonding electron pairs on the central atom do, however,

142   Chapter 11   Chemical Bonding

cause minor changes in the bond angles, compared to the ideal regular geometric structure.

76. You will remember from high school geometry, that two points in space are all that is needed to define a straight line. A diatomic molecule represents two points (the nuclei of the atoms) in space.

78. In NF_3, the nitrogen atom has *four* pairs of valence electrons, whereas in BF_3, there are only *three* pairs of valence electrons around the boron atom. The nonbonding electron pair on nitrogen in NF_3 pushes the three F atoms out of the plane of the N atom.

80. Each of the indicated atoms is surrounded by *four pairs* of electrons with a *tetrahedral* orientation. Although the electron pairs are arranged tetrahedrally, realize that the overall geometric shape of the molecules may *not* be tetrahedral, depending on what atoms are *bonded* to the four pairs of electrons on the central atom.

82. a.   tetrahedral (four electron pairs on C, and four atoms attached)

   b.   nonlinear, *V*-shaped (four electron pairs on S, but only two atoms attached)

   c.   tetrahedral (four electron pairs on Ge, and four atoms attached)

84. a.   basically tetrahedral around the P atom (the hydrogen atoms are attached to two of the oxygen atoms and do not affect greatly the geometrical arrangement of the oxygen atoms around the phosphorus)

   b.   tetrahedral (4 electron pairs on Cl, and 4 atoms attached)

   c.   trigonal pyramidal (4 electron pairs on S, and 3 atoms attached)

86. a.   approximately 109.5° (the molecule is *V*-shaped or nonlinear)

   b.   approximately 109.5° (the molecule is trigonal pyramidal)

   c.   109.5°

   d.   approximately 120° (the double bond makes the molecule flat)

88. Acetylene is a linear molecule with bond angles of 180°.

90. double

92. The bond with the larger electronegativity difference will be the more polar bond. See Figure 11.3 for electronegativities.

   a.   S-F

   b.   P-O

   c.   C-H

94. The bond energy of a chemical bond is the quantity of energy required to break the bond and separate the atoms.

96. In each case, the element *higher up* within a group of the periodic table has the higher electronegativity.

   a. Be
   b. N
   c. F

98. For a bond to be polar covalent, the atoms involved in the bond must have different electronegativities (must be of different elements).

   a. polar covalent (different elements)
   b. *non*polar covalent (two atoms of the same element)
   c. polar covalent (different elements)
   d. *non*polar covalent (atoms of the same element)

100. In a diatomic molecule containing two different elements, the more electronegative atom will be the negative end of the molecule, and the *less* electronegative atom will be the positive end.

   a. oxygen
   b. bromine
   c. iodine

102. a. Al $1s^2\ 2s^2\ 2p^6\ 3s^2\ 3p^1$

   Al^{3+} $1s^2\ 2s^2\ 2p^6$

   Ne has the same configuration as Al^{3+}.

   b. Br $1s^2\ 2s^2\ 2p^6\ 3s^2\ 3p^6\ 4s^2\ 3d^{10}\ 4p^5$

   Br$^-$ $1s^2\ 2s^2\ 2p^6\ 3s^2\ 3p^6\ 4s^2\ 3d^{10}\ 4p^6$

   Kr has the same configuration as Br$^-$.

   c. Ca $1s^2\ 2s^2\ 2p^6\ 3s^2\ 3p^6\ 4s^2$

   Ca^{2+} $1s^2\ 2s^2\ 2p^6\ 3s^2\ 3p^6$

   Ar has the same configuration as Ca^{2+}.

   d. Li $1s^2\ 2s^1$

   Li$^+$ $1s^2$

   He has the same configuration as Li$^+$.

144 Chapter 11   Chemical Bonding

    e.   F $1s^2\ 2s^2\ 2p^5$

           F^- $1s^2\ 2s^2\ 2p^6$

        Ne has the same configuration as F^-.

104. a. Na_2Se Na has one electron more than a noble gas; Se has two electrons fewer than a noble gas.

     b.   RbF   Rb has one electron more than a noble gas; F has one electron less than a noble gas.

     c. K_2Te K has one electron more than a noble gas; Te has two electrons fewer than a noble gas.

     d.   BaSe   Ba has two electrons more than a noble gas; Se has two electrons fewer than a noble gas.

     e.   KAt   K has one electron more than a noble gas; At has one electron less than a noble gas.

     f.   FrCl   Fr has one electron more than a noble gas; Cl has one electron less than a noble gas.

106. Relative ionic sizes are indicated in Figure 11.9.

     a. Na^+

     b. Al^{3+}

     c. F^-

     d. Na^+

108. a.   H provides 1; N provides 5; each O provides 6; total valence electrons = 24

     b.   each H provides 1; S provides 6; each O provides 6; total valence electrons = 32

     c.   each H provides 1; P provides 5; each O provides 6; total valence electrons = 32

     d.   H provides 1; Cl provides 7; each O provides 6; total valence electrons = 32

110. a. N_2H_4 Each N provides 5 valence electrons.
                       Each H provides 1 valence electron.
                       Total valence electrons = 14

$$H-\overset{..}{\underset{H}{N}}-\overset{..}{\underset{H}{N}}-H$$

b. C₂H₆   Each C provides 4 valence electrons.
Each H provides 1 valence electron.
Total valence electrons = 14

$$\begin{array}{c} \text{H} \quad \text{H} \\ | \quad\; | \\ \text{H}-\text{C}-\text{C}-\text{H} \\ | \quad\; | \\ \text{H} \quad \text{H} \end{array}$$

c. NCl₃   N provides 5 valence electrons.
Each Cl provides 7 valence electrons.
Total valence electrons = 26

$$\begin{array}{c} :\!\ddot{\text{C}\text{l}}-\ddot{\text{N}}-\ddot{\text{C}\text{l}}\!: \\ | \\ :\!\ddot{\text{C}\text{l}}\!: \end{array}$$

d. SiCl₄   Si provides 4 valence electrons.
Each Cl provides 7 valence electrons.
Total valence electrons = 32

$$\begin{array}{c} :\!\ddot{\text{C}\text{l}}\!: \\ | \\ :\!\ddot{\text{C}\text{l}}-\text{Si}-\ddot{\text{C}\text{l}}\!: \\ | \\ :\!\ddot{\text{C}\text{l}}\!: \end{array}$$

112. a. NO₃⁻   N provides 5 valence electrons.
Each O provides 6 valence electrons.
The 1− charge means one additional valence electron.
Total valence electrons = 24

$$\left[\begin{array}{c} \ddot{\text{O}}=\text{N}-\ddot{\text{O}}: \\ | \\ :\ddot{\text{O}}: \end{array}\right]^{-} \leftrightarrow \left[\begin{array}{c} :\ddot{\text{O}}-\text{N}=\ddot{\text{O}} \\ | \\ :\ddot{\text{O}}: \end{array}\right]^{-} \leftrightarrow \left[\begin{array}{c} :\ddot{\text{O}}-\text{N}-\ddot{\text{O}}: \\ \| \\ :\text{O}: \end{array}\right]^{-}$$

b. CO₃²⁻   C provides 4 valence electrons.
Each O provides 6 valence electrons.
The 2− charge means two additional valence electrons.
Total valence electrons = 24

$$\left[\begin{array}{c} \ddot{\text{O}}=\text{C}-\ddot{\text{O}}: \\ | \\ :\ddot{\text{O}}: \end{array}\right]^{2-} \leftrightarrow \left[\begin{array}{c} :\ddot{\text{O}}-\text{C}=\ddot{\text{O}} \\ | \\ :\ddot{\text{O}}: \end{array}\right]^{2-} \leftrightarrow \left[\begin{array}{c} :\ddot{\text{O}}-\text{C}-\ddot{\text{O}}: \\ \| \\ :\text{O}: \end{array}\right]^{2-}$$

c. NH_4^+ N provides 5 valence electrons.
Each H provides 1 valence electron.
The 1+ charge means one *less* valence electron.
Total valence electrons = 8

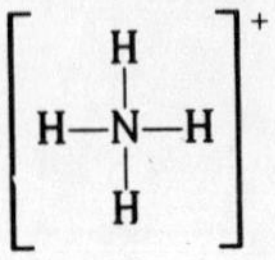

$$\left[\begin{array}{c} H \\ | \\ H-N-H \\ | \\ H \end{array}\right]^+$$

114. a. four electron pairs arranged tetrahedrally about C

   b. four electron pairs arranged tetrahedrally about Ge

   c. three electron pairs arranged trigonally (planar) around B

116. a. ClO_3^-, trigonal pyramid (lone pair on Cl)

   b. ClO_2^-, nonlinear (*V*-shaped, two lone pairs on Cl)

   c. ClO_4^-, tetrahedral (all pairs on Cl are bonding)

118. a. nonlinear (V-shaped)

   b. trigonal planar

   c. basically trigonal planar around the C (the H is attached to one of the O atoms, and distorts the shape around the carbon only slightly)

   d. linear

120. Ionic compounds tend to be hard, crystalline substances with relatively high melting and boiling points. Covalently bonded substances tend to be gases, liquids, or relatively soft solids, with much lower melting and boiling points.

Cumulative Review: Chapters 10 and 11

2. An atom is said to be in its ground state when it is in its lowest possible energy state. When an atom possesses more energy than in its ground state, the atom is said to be in an excited state. An atom is promoted from its ground state to an excited state by absorbing energy; when the atom returns from an excited state to its ground state it emits the excess energy as electromagnetic radiation. Atoms do not gain or emit radiation randomly, but rather do so only in discrete bundles of radiation called photons. The photons of radiation emitted by atoms are characterized by the wavelength (color) of the radiation: longer wavelength photons carry less energy than shorter wavelength photons. The energy of a photon emitted by an atom corresponds exactly to the difference in energy between two allowed energy states in an atom: thus, we can use an observable phenomenon (emission of light by excited atoms), to gain insight into the energy changes taking place within the atom.

4. Bohr pictured the electron moving in only certain circular orbits around the nucleus. Each particular orbit (corresponding to a particular distance from the nucleus) had associated with it a particular energy (resulting from the attraction between the nucleus and the electron). When an atom absorbs energy, the electron moves from its ground state in the orbit closest to the nucleus ($n = 1$) to an orbit farther away from the nucleus ($n = 2, 3, 4, ...$). When an excited atom returns to its ground state, corresponding to the electron moving from an outer orbit to the orbit nearest the nucleus, the atom emits the excess energy as radiation. Since the Bohr orbits are of fixed distances from the nucleus and from each other, when an electron moves from one fixed orbit to another, the energy change is of a definite amount, which corresponds to a photon being emitted of a particular characteristic wavelength and energy. The original Bohr theory worked very well for hydrogen: Bohr even predicted emission wavelengths for hydrogen which had not yet been seen, which were subsequently found at the exact wavelengths Bohr had calculated. However, when the simple Bohr model for the atom was applied to the emission spectra of other elements, the theory could not predict or explain the observed emission spectra.

6. The lowest energy hydrogen atomic orbital is called the 1s orbital. The 1s orbital is spherical in shape (that is, the electron density around the nucleus is uniform in all directions from the nucleus). The 1s orbital represents a probability map of electron density around the nucleus for the first principal energy level. The orbital does not have a sharp edge (it appears fuzzy) since the probability of finding the electron does not drop off completely suddenly with distance from the nucleus. The orbital does not represent just a spherical surface on which the electron moves (this would be similar to Bohr's original theory): when we draw a picture to represent the 1s orbital we are indicating that the probability of finding the electron within this region of space is greater than 90%. We know that the likelihood of finding the electron within this orbital is very high, but we still

don't know exactly where in this region the electron is at a given instant in time.

8. The third principal energy level of hydrogen is divided into three sublevels, the $3s$, $3p$, and $3d$ sublevels. The $3s$ subshell consists of the single $3s$ orbital: like the other s orbitals, the $3s$ orbital is spherical in shape. The $3p$ subshell consists of a set of three equal-energy $3p$ orbitals: each of these $3p$ orbitals has the same shape ("dumbbell"), but each of the $3p$ orbitals is oriented in a different direction in space. The $3d$ subshell consists of a set of five $3d$ orbitals: the $3d$ orbitals have the shapes indicated in Figure 10.23, and are oriented in different directions around the nucleus (students sometimes say that the $3d$ orbitals have the shape of a 4-leaf clover). The fourth principal energy level of hydrogen is dived into four sublevels, the $4s$, $4p$, $4d$, and $4f$ orbitals. The $4s$ subshell consists of the single $4s$ orbital. The $4p$ subshell consists of a set of three $4p$ orbitals. The $4d$ subshell consists of a set of five $4d$ orbitals. The shapes of the $4s$, $4p$, and $4d$ orbitals are the same as the shapes of the orbitals of the third principal energy level (the orbitals of the fourth principal energy level are larger and further from the nucleus than the orbitals of the third level, however). The fourth principal energy level, because it is further from the nucleus, also contains a $4f$ subshell, consisting of seven $4f$ orbitals (the shapes of the $4f$ orbitals are beyond the scope of this text).

10. Atoms have a series of principal energy levels symbolized by the letter n. The $n = 1$ level is the closest to the nucleus, and the energies of the levels increase as the value of n increases going out from the nucleus. Each principal energy level is divided into a set of sublevels of different characteristic shapes (designated by the letters s, p, d, and f). Each sublevel is further subdivided into a set of orbitals: each s subshell consists of a single s orbital; each p subshell consists of a set of three p orbitals; each d subshell consists of a set of five d orbitals; etc. A given orbital can be empty or it can contain one or two electrons, but never more than two electrons (if an orbital contains two electrons, then the electrons must have opposite intrinsic spins). The shape we picture for an orbital represents only a probability map for finding electrons: the shape does not represent a trajectory or pathway for electron movements.

12. The valence electrons of an atoms are the electrons in the outermost shell of the atom. The core electrons are those in principal energy levels closer to the nucleus than the outermost shell (the core electrons are the electrons that are not valence electrons). Since the valence electrons are those in the outermost shell of the atom, it is these electrons that are affected by the presence of other atoms, and which are gained, lost, or shared with other atoms. The periodic table is basically arranged in terms of the valence electronic configurations of the elements (elements in the same vertical group have similar configurations): for example, Group 1 is the first column in the periodic table because all the elements in Group 1 have one valence electron.

14. The general periodic table you drew for Question 13 should be similar to that found in Figure 10.27 of the text. Just from the column and row location of an element, you should be able to determine what the valence shell of an element has for its electronic configuration. For example, the element in the third horizontal row, in the second vertical column, has $3s^2$ as its valence configuration. We know that the valence electrons are in the $n = 3$ shell because the element is in the third horizontal row. We know that the valence electrons are s electrons because the first two electrons in a horizontal row are always in an s subshell. We know that there are two electrons because the element is the second element in the horizontal row. As an additional example, the element in the seventh vertical column of the second horizontal row in the periodic table has valence configuration $2s^2 2p^5$.

16. The ionization energy of an atom represents the energy required to remove an electron from the atom. As one goes from top to bottom in a vertical group in the periodic table, the ionization energies decrease (it becomes easier to remove an electron). As one goes down within a group, the valence electrons are farther and farther from the nucleus and are less tightly held. The ionization energies increase when going from left to right within a horizontal row within the periodic table. The left-hand side of the periodic table is where the metallic elements are found, which lose electrons relatively easily. The right-hand side of the periodic table is where the nonmetallic elements are found: rather than losing electrons, these elements tend to gain electrons. Within a given horizontal row in the periodic table, the valence electrons are all in the same principal energy shell: however, as you go from left to right in the horizontal row, the nuclear charge which holds onto the electrons is increasing one unit with each successive element, making it that much more difficult to remove an electron. The relative sizes of atoms also vary systematically with the location of an element in the periodic table. Within a given vertical group, the atoms get progressively larger when going from the top of the group to the bottom: the valence electrons of the atoms are in progressively higher principal energy shells (and are progressively further from the nucleus) as we go down in a group. In going from left to right within a horizontal row in the periodic table, the atoms get progressively smaller. Although all the elements in a given horizontal row in the periodic table have their valence electrons in the same principal energy shell, the nuclear charge is progressively increasing from left to right, making the given valence shell progressively smaller as the electrons are drawn more closely to the nucleus.

18. Ionic bonding results when elements of very different electro-negativities react with each other. Typically a metallic element reacts with a nonmetallic element, with the metallic element losing electrons and forming forming positive ions and the nonmetallic element gaining electrons and forming negative ions. Sodium chloride, NaCl, is an example of a typical ionic compound. The aggregate form of such a compound consists of a crystal lattice of alternating positively and negatively charged ions. A given positive ion is attracted by several surrounding negatively charged ions, and a given negative ion is

attracted by several surrounding positively charged ions. Similar electrostatic attractions go on in three dimensions throughout the crystal of ionic solid, leading to a very stable system (with very high melting and boiling points, for example). We know that ionically bonded solids do not conduct electricity in the solid state (since the ions are held tightly in place by all the attractive forces), but such substances are strong electrolytes when melted or when dissolved in water (either of which process sets the ions free to move around).

20. Electronegativity represents the relative ability of an atom in a molecule to attract shared electrons towards itself. In order for a bond to be polar, one of the atoms in the bond must attract the shared electron pair towards itself and away from the other atom of the bond: this can only happen if one atom of the bond is more electronegative than the other (that is, that there is a considerable difference in electronegativity for the two atoms of the bond). If two atoms in a bond have the same electronegativity, then the two atoms pull the electron pair equally and the bond is nonpolar covalent. If two atoms sharing a pair of electrons have vastly different electronegativities, the electron pair will be pulled so strongly by the more electronegative atom that a negative ion may be formed (as well as a positive ion for the second atom) and ionic bonding will result. If the difference in electronegativity between two atoms sharing an electron pair is somewhere in between these two extremes (equal sharing of the electron pair and formation of ions), then a polar covalent bond results.

22. It has been observed over many, many experiments that when an active metal like sodium or magnesium reacts with a nonmetal, the sodium atoms always form Na^+ ions and the magnesium atoms always form Mg^{2+} ions. It has been further observed that aluminum always forms only the Al^{3+} ion, and that when nitrogen, oxygen, or fluorine form simple ions, the ions that are formed are always N^{3-}, O^{2-}, and F^-, respectively. Clearly the facts that these elements always form the same ions and that those ions all contain eight electrons in the outermost shell, led scientists to speculate that there must be something very fundamentally stable about a species that has eight electrons in its outermost shell (like the noble gas neon). The repeated observation that so many elements, when reacting, tend to attain an electronic configuration that is isoelectronic with a noble gas led chemists to speculate that all elements try to attain such a configuration for their outermost shells. In general, when atoms of a metal react with atoms of a nonmetal, the metal atoms lose electrons until they have the configuration of the preceding noble gas, and the nonmetal atoms gain electrons until they have the configuration of the following noble gas. Covalently and polar covalently bonded molecules also strive to attain pseudo-noble gas electronic configurations. For a covalently bonded molecule like F_2, in which neither fluorine atom has a greater tendency than the other to gain or lose electrons completely, each F atom provides one electron of the pair of electrons which constitutes the covalent bond. Each F atom feels also the influence of the other F atom's electron in the shared pair, and each F atom effectively fills its outermost shell. Similarly, in polar covalently bonded molecules like HF or HCl, the shared pair of

24. Bonding between atoms to form a molecule involves only the valence electrons of the atoms (not the inner core electrons). So when we draw the Lewis structure of a molecule, we show only these valence electrons (both bonding valence electrons and non-bonding valence electrons, however). The most important requisite for the formation of a stable compound, and which we try to demonstrate when we write Lewis structures, is that each atom of a molecule attains a noble gas electron configuration. When we write Lewis structures, we arrange the bonding and nonbonding valence electrons to try to complete the octet (or duet) for as many atoms as is possible.

26. Obviously, you could choose practically any molecules for your discussion. Let's illustrate the method for ammonia, NH_3. First count up the total number of valence electrons available in the molecule (without regard to what atom they officially come from); remember that for the representative elements, the number of valence electrons is indicated by what group the element is found in on the periodic table. For NH_3, since nitrogen is in Group 5, one nitrogen atom would contribute five valence electron. Since hydrogen atoms only have one electron each, the three hydrogen atoms provide an additional three valence electrons, for a total of eight valence electrons overall. Next write down the symbols for the atoms in the molecule, and use one pair of electrons (represented by a line) to form a bond between each pair of bound atoms.

$$\begin{array}{c} H-N-H \\ | \\ H \end{array}$$

These three bonds use six of the eight valence electrons. Since each hydrogen already has its duet in what we have drawn so far, while the nitrogen atom only has six electrons around it so far, the final two valence electrons must represent a lone pair on the nitrogen.

$$\begin{array}{c} H-\ddot{N}-H \\ | \\ H \end{array}$$

28. There are several types of exceptions to the octet rule described in the text. The octet rule is really a "rule of thumb" which we apply to molecules unless we have some evidence that a molecule does not follow the rule. There are some common molecules which, from experimental measurements, we know do not follow the octet rule. Boron and beryllium compounds sometimes do not fit the octet rule. For example, in BF_3, the boron atom only has six valence electrons, whereas in BeF_2, the

beryllium atom only has four valence electrons. Other molecules which are exceptions to the octet rule include any molecule with an odd number of valence electrons (such as NO or NO_2): you can't get an octet (an even number) of electrons around each atom in a molecule with an odd number of valence electrons. Even the oxygen gas we breathe is an exception to the octet rule: although we can write a Lewis structure for O_2 satisfying the octet rule for each oxygen, we know from experiment that O_2 contains unpaired electrons (which would not be consistent with a structure in which all the electrons were paired up.)

30.

Number of valence pairs	bond angle	example(s)
2	180°	BeF_2, BeH_2
3	120°	BCl_3
4	109.5°	CH_4, CCl_4, GeF_4

Chapter 12   Gases

2. Solids and liquids have essentially fixed volumes and are not able to be compressed easily. Gases have volumes that depend on their conditions, and can be compressed or expanded by changes in those conditions. Although the particles of matter in solids are essentially fixed in position (the solid is rigid), the particles in liquids and gases are free to move.

4. Figure 12.2 in the text shows a simple mercury barometer: a tube filled with mercury is inverted over a reservoir containing mercury which is open to the atmosphere. When the tube is inverted, the mercury falls to a level at which the pressure of the atmosphere is sufficient to support the column of mercury. One standard atmosphere of pressure is taken to be the pressure capable of supporting a column of mercury to a height of 760.0 mm above the reservoir level.

6. 760 (defined quantity)

8. a. $105.2 \text{ kPa} \times \dfrac{1 \text{ atm}}{101.325 \text{ kPa}} = 1.038 \text{ atm}$

    b. $75.2 \text{ cm Hg} \times \dfrac{10 \text{ mm Hg}}{1 \text{ cm Hg}} \times \dfrac{1 \text{ atm}}{760 \text{ mm Hg}} = 0.989 \text{ atm}$

    c. $752 \text{ mm Hg} \times \dfrac{1 \text{ atm}}{760 \text{ mm Hg}} = 0.989 \text{ atm}$

    d. 767 torr = 767 mm Hg

    $767 \text{ torr} \times \dfrac{1 \text{ atm}}{760 \text{ torr}} = 1.01 \text{ atm}$

10. a. $0.9975 \text{ atm} \times \dfrac{760 \text{ mm Hg}}{1 \text{ atm}} = 758.1 \text{ mm Hg}$

    b. $225{,}400 \text{ Pa} \times \dfrac{760 \text{ mm Hg}}{101{,}325 \text{ Pa}} = 1691 \text{ mm Hg}$

    c. $99.7 \text{ kPa} \times \dfrac{760 \text{ mm Hg}}{101.325 \text{ kPa}} = 748 \text{ mm Hg}$

    d. $1.078 \text{ atm} \times \dfrac{760 \text{ mm Hg}}{1 \text{ atm}} = 819.3 \text{ mm Hg}$

12. a. $2.07 \times 10^6 \text{ Pa} \times \dfrac{1 \text{ kPa}}{10^3 \text{ Pa}} = 2.07 \times 10^3 \text{ kPa}$

    b. $795 \text{ mm Hg} \times \dfrac{101{,}325 \text{ Pa}}{760 \text{ mm Hg}} \times \dfrac{1 \text{ kPa}}{10^3 \text{ Pa}} = 106 \text{ kPa}$

    c. $10.9 \text{ atm} \times \dfrac{101{,}325 \text{ Pa}}{1 \text{ atm}} \times \dfrac{1 \text{ kPa}}{10^3 \text{ Pa}} = 1.10 \times 10^3 \text{ kPa}$

154   Chapter 12   Gases

d. $659 \text{ torr} \times \dfrac{101,325 \text{ Pa}}{760 \text{ torr}} \times \dfrac{1 \text{ kPa}}{10^3 \text{ Pa}} = 87.9 \text{ kPa}$

14. increases

16. $PV = k$; $P_1V_1 = P_2V_2$

18. a. $P_1 = 1.00$ atm $P_2 = 699$ torr $= 0.920$ atm
       $V_1 = 541$ mL $V_2 = ?$ mL

       $V_2 = \dfrac{P_1V_1}{P_2} = \dfrac{(1.00 \text{ atm})(541 \text{ mL})}{(0.920 \text{ atm})} = 588$ mL

    b. $P_1 = 110.2$ kPa $P_2 = 0.995$ atm $= 100.8$ kPa
       $V_1 = 2.32$ L $V_2 = ?$ L

       $V_2 = \dfrac{P_1V_1}{P_2} = \dfrac{(110.2 \text{ kPa})(2.32 \text{ L})}{(100.8 \text{ kPa})} = 2.54$ L

    c. $P_1 = 135$ atm $= 1.026 \times 10^5$ mm Hg $P_2 = ?$ mm Hg
       $V_1 = 4.15$ mL $V_2 = 10.0$ mL

       $P_2 = \dfrac{P_1V_1}{V_2} = \dfrac{(1.026 \times 10^5 \text{ mm Hg})(4.15 \text{ mL})}{(10.0 \text{ mL})} = 4.26 \times 10^4$ mm Hg

20. a. $P_1 = 1.07$ atm $P_2 = 2.14$ atm
       $V_1 = 291$ mL $V_2 = ?$ mL

       $V_2 = \dfrac{P_1V_1}{P_2} = \dfrac{(1.07 \text{ atm})(291 \text{ mL})}{(2.14 \text{ atm})} = 146$ mL

    b. $P_1 = 755$ mm Hg $P_2 = 3.51$ atm $= 2668$ mm Hg
       $V_1 = 1.25$ L $V_2 = ?$ L

       $V_2 = \dfrac{P_1V_1}{P_2} = \dfrac{(755 \text{ mm Hg})(1.25 \text{ L})}{(2668 \text{ mm Hg})} = 0.354$ L

    c. $P_1 = 101.4$ kPa $= 760.6$ mm Hg $P_2 = ?$ mm Hg
       $V_1 = 2.71$ L $V_2 = 3.00$ L

       $P_2 = \dfrac{P_1V_1}{V_2} = \dfrac{(760.6 \text{ mm Hg})(2.71 \text{ L})}{(3.00 \text{ L})} = 687$ mm Hg

22. If the pressure exerted on the gas in the balloon is decreased, the volume of the gas in the balloon will increase in inverse proportion to the factor by which the pressure was changed. The factor in this example is (1.01 atm/0.562 atm) = 1.80

24. $P_1 = 760$ mm $= 1.00$ atm $P_2 = ?$ atm
    $V_1 = 1.00$ L $V_2 = 50.0$ mL $= 0.0500$ L

$$P_2 = \frac{P_1 V_1}{V_2} = \frac{(1.00 \text{ atm})(1.00 \text{ L})}{(0.0500 \text{ L})} = 20.0 \text{ atm}$$

26. Charles's law states that the volume of an ideal gas sample varies linearly with the absolute temperature of the gas sample. An experiment such as those indicated in Figures 12.7 and 12.8 can be performed to determine absolute zero. The volume of a sample of gas is measured at several convenient temperatures (e.g., between 0° and 100° C) and the data is plotted. The straight line obtained is then extrapolated to the point where the volume of the gas would become zero. The temperature at which the volume of the gas would be predicted to become zero is then absolute zero.

28. $V = bT$; $V_1/T_1 = V_2/T_2$

30. $V_1 = 525$ mL $\qquad V_2 = ?$ mL

    $T_1 = 25$ °C $= 298$ K $\qquad T_2 = 50$ °C $= 323$ K

    $$V_2 = \frac{V_1 T_2}{T_1} = \frac{(525 \text{ mL})(323 \text{ K})}{(298 \text{ K})} = 569 \text{ mL}$$

32. a. $V_1 = 25.0$ L $\qquad V_2 = 50.0$ L

    $T_1 = 0$ °C $= 273$ K $\qquad T_2 = ?$ °C

    $$T_2 = \frac{V_2 T_1}{V_1} = \frac{(50.0 \text{ L})(273 \text{ K})}{(25.0 \text{ L})} = 546 \text{ K} = 273 \text{ °C}$$

    b. $V_1 = 247$ mL $\qquad V_2 = 255$ mL

    $T_1 = 25$ °C $= 298$ K $\qquad T_2 = ?$ °C

    $$T_2 = \frac{V_2 T_1}{V_1} = \frac{(255 \text{ mL})(298 \text{ K})}{(247 \text{ mL})} = 308 \text{ K} = 35 \text{ °C}$$

    c. $V_1 = 1.00$ mL $\qquad V_2 = ?$ mL

    $T_1 = -272$ °C $= 1$ K $\qquad T_2 = 25$ °C $= 298$ K

    $$V_2 = \frac{V_1 T_2}{T_1} = \frac{(1.00 \text{ mL})(298 \text{ K})}{(1 \text{ K})} = 298 \text{ mL}$$

34. a. $V_1 = 2.01 \times 10^2$ L $\qquad V_2 = 5.00$ L

    $T_1 = 1150$ °C $= 1423$ K $\qquad T_2 = ?$ °C

    $$T_2 = \frac{V_2 T_1}{V_1} = \frac{(5.00 \text{ L})(1423 \text{ K})}{(2.01 \times 10^2 \text{ L})} = 35.4 \text{ K} = -238 \text{ °C}$$

156   Chapter 12   Gases

b. $V_1 = 44.2$ mL $V_2 = ?$ mL

     $T_1 = 298$ K $T_2 = 0$ K

$$V_2 = \frac{V_1 T_2}{T_1} = \frac{(44.2 \text{ mL})(0 \text{ K})}{(298 \text{ K})} = 0 \text{ mL} \quad (0 \text{ K is absolute zero})$$

c. $V_1 = 44.2$ mL $V_2 = ?$ mL

     $T_1 = 298$ K $T_2 = 0$ °C $= 273$ K

$$V_2 = \frac{V_1 T_2}{T_1} = \frac{(44.2 \text{ mL})(273 \text{ K})}{(298 \text{ K})} = 40.5 \text{ mL}$$

36. You should be able to answer these without having to set up a formal calculation. Charles's law says that the volume of a gas sample is *directly* proportional to its absolute temperature. So if a sample of neon has a volume of 266 mL at 25.2 °C (298 K), then the volume will become half as big at half the absolute temperature (149 K, -124 °C). The volume of the gas sample will become twice as big at twice the absolute temperature (596 K, 323°C).

38. One method of solution (using Charles's Law in the form $V_1/T_1 = V_2/T_2$) is shown in Example 12.6. A second method might make use of Charles's Law in the form $V = kT$. Using the information that the gas thermometer has a volume of 135 mL at 11°C (284 K), we can solve for the value of the proportionality constant k in the formula, and then use this information to calculate the additional temperatures requested.

$V = kT$

$$k = \frac{V}{T} = \frac{135 \text{ mL}}{284 \text{ K}} = 0.4754 \text{ mL/K}$$

$$T = \frac{V}{k} = \frac{V}{0.475 \text{ mL/K}} = 2.104V$$

For 113 mL, $T = 2.104(113) = 238$ K (-35°C)
For 142 mL, $T = 2.104(142) = 299$ K (26°C)
For 155 mL, $T = 2.104(155) = 326$ K (53°C)
For 127 mL, $T = 2.104(127) = 267$ K (-6°C)

40. $V = an;$ $V_1/n_1 = V_2/n_2$

42. $V_1 = 12.0$ L $V_2 = ?$ L

      $n_1 = 2.01$ g $= 0.502$ mol $n_2 = 6.52$ g $= 1.63$ mol

$$V_2 = \frac{V_1 n_2}{n_1} = \frac{(12.0 \text{ L})(1.63 \text{ mol})}{(0.502 \text{ L})} = 39.0 \text{ L}$$

44. $V_1 = 100.$ L $V_2 = ?$ L

      $n_1 = 46.2$ g/32.00 g mol^{-1} $n_2 = 5.00$ g/32.00 g mol^{-1}

$$V_2 = \frac{V_1 n_2}{n_1} = \frac{(100.\text{ L})(5.00 \text{ g}/32.00 \text{ g mol}^{-1})}{(46.2 \text{ g}/32.00 \text{ g mol}^{-1})} = 10.8 \text{ L}$$

Note that the *molar mass* of the O_2 gas cancels out in this calculation. Since the number of moles of Ne (or any gas) present in a sample is *directly proportional* to the mass of the gas sample, the problem could also have been set up directly in terms of the masses.

46. Real gases most closely approach ideal gas behavior under conditions of relatively high temperatures (0°C or higher) and relatively low pressures (1 atm or lower).

48. For an ideal gas, $PV = nRT$ is true under any conditions. Consider a particular sample of gas (so that n remains constant) at a particular fixed pressure (so that P remains constant also). Suppose that at temperature T_1 the volume of the gas sample is V_1. Then for this set of conditions, the ideal gas equation would be given by

    $$PV_1 = nRT_1$$

    If we then change the temperature of the gas sample to a new temperature T_2, the volume of the gas sample changes to a new volume V_2. For this new set of conditions, the ideal gas equation would be given by

    $$PV_2 = nRT_2$$

    If we make a ratio of these two expressions for the ideal gas equation for this gas sample, and cancel out terms that are constant for this situation (P, n, and R) we get

    $$\frac{PV_1}{PV_2} = \frac{nRT_1}{nRT_2}$$

    $$\frac{V_1}{V_2} = \frac{T_1}{T_2}$$

    which can be rearranged to the familiar form of Charles's law

    $$\frac{V_1}{T_1} = \frac{V_2}{T_2}$$

50. a. $P = 782$ mm Hg $= 1.03$ atm

    $T = 27$ °C $= 300$ K

    $$V = \frac{nRT}{P} = \frac{(0.210 \text{ mol})(0.08206 \text{ L atm mol}^{-1} \text{ K}^{-1})(300 \text{ K})}{(1.03 \text{ atm})} = 5.02 \text{ L}$$

    b. $V = 644$ mL $= 0.644$ L

    $$P = \frac{nRT}{V} = \frac{(0.0921 \text{ mol})(0.08206 \text{ L atm mol}^{-1} \text{ K}^{-1})(303 \text{ K})}{(0.644 \text{ L})} = 3.56 \text{ atm}$$

158 Chapter 12 Gases

$$P = 3.56 \text{ atm} = 2.70 \times 10^3 \text{ mm Hg}$$

c. $P = 745 \text{ mm} = 0.980 \text{ atm}$

$$T = \frac{PV}{nR} = \frac{(0.980 \text{ atm})(11.2 \text{ L})}{(0.401 \text{ mol})(0.08206 \text{ L atm mol}^{-1} \text{ K}^{-1})} = 334 \text{ K}$$

52. a. $T = 25 \text{ °C} = 298 \text{ K}$

$$V = \frac{(0.00831 \text{ mol})(0.08206 \text{ L atm mol}^{-1} \text{ K}^{-1})(298 \text{ K})}{(1.01 \text{ atm})} = 0.201 \text{ L}$$

b. $V = 602 \text{ mL} = 0.602 \text{ L}$

$$P = \frac{(8.01 \times 10^{-3} \text{ mol})(0.08206 \text{ L atm mol}^{-1} \text{ K}^{-1})(310 \text{ K})}{(0.602 \text{ L})} = 0.338 \text{ atm}$$

c. $V = 629 \text{ mL} = 0.629 \text{ L}$

$T = 35 \text{ °C} = 308 \text{ K}$

$$n = \frac{(0.998 \text{ atm})(0.629 \text{ L})}{(0.08206 \text{ L atm mol}^{-1} \text{ K}^{-1})(308 \text{ K})} = 2.48 \times 10^{-2} \text{ mol}$$

54. Molar mass of O_2 = 32.00 g

$$6.21 \text{ g } O_2 \times \frac{1 \text{ mol}}{32.00 \text{ g}} = 0.194 \text{ mol } O_2$$

$$T = \frac{PV}{nR} = \frac{(5.00 \text{ atm})(10.0 \text{ L})}{(0.194 \text{ mol})(0.08206 \text{ L atm mol}^{-1} \text{ K}^{-1})} = 3140 \text{ K}$$

56. The number of moles of *any* ideal gas that can be contained in the tank under the given conditions can first be calculated.

$T = 24 \text{ °C} = 297 \text{ K}$

$$n = \frac{PV}{RT} = \frac{(135 \text{ atm})(200 \text{ L})}{(0.08206 \text{ L atm mol}^{-1} \text{ K}^{-1})(297 \text{ K})} = 1.11 \times 10^3 \text{ mol gas}$$

Molar masses: He, 4.003 g; H_2, 2.016 g

for He: $1.11 \times 10^3 \text{ mol He} \times \frac{4.003 \text{ g}}{1 \text{ mol}} = 4.44 \times 10^3 \text{ g He} = 4.44 \text{ kg He}$

for H_2: $1.11 \times 10^3 \text{ mol } H_2 \times \frac{2.016 \text{ g}}{1 \text{ mol}} = 2.24 \times 10^3 \text{ g } H_2 = 2.24 \text{ kg } H_2$

58. Molar mass of N_2 = 28.02 g

$$16.3 \text{ g } N_2 \times \frac{1 \text{ mol}}{28.02 \text{ g}} = 0.582 \text{ mol } N_2$$

$$T = \frac{PV}{nR} = \frac{(1.25 \text{ atm})(25.0 \text{ L})}{(0.582 \text{ mol})(0.08206 \text{ L atm mol}^{-1} \text{ K}^{-1})} = 654 \text{ K} = 381 \text{ °C}$$

Chapter 12   Gases   159

60. Molar mass of O_2 = 32.00 g

56.2 kg = 5.62 × 10⁴ g

$$5.62 \times 10^4 \text{ g} \times \frac{1 \text{ mol}}{32.00 \text{ g}} = 1.76 \times 10^3 \text{ mol}$$

T = 21 °C = 294 K

$$P = \frac{nRT}{V} = \frac{(1.76 \times 10^3 \text{ mol})(0.08206 \text{ L atm mol}^{-1} \text{ K}^{-1})(294 \text{ K})}{(125 \text{ L})}$$

P = 340 atm

62. P_1 = 0.981 atm P_2 = 1.15 atm
    V_1 = 125 mL V_2 = ? mL
    T_1 = 100 °C = 373 K T_2 = 25 °C = 298 K

$$V_2 = \frac{P_1 V_1 T_2}{P_2 T_1} = \frac{(0.981 \text{ atm})(125 \text{ mL})(298 \text{ K})}{(1.15 \text{ atm})(373 \text{ K})} = 85.2 \text{ mL}$$

64. Molar mass of H_2 = 2.016 g

$$5.00 \text{ g } H_2 \times \frac{1 \text{ mol}}{2.016 \text{ g}} = 2.48 \text{ mol } H_2$$

P = 761 mm = 1.001 atm

$$T = \frac{PV}{nR} = \frac{(1.001 \text{ atm})(50.0 \text{ L})}{(2.48 \text{ mol})(0.08206 \text{ L atm mol}^{-1} \text{ K}^{-1})} = 246 \text{ K} = -27 \text{ °C}$$

66. As a gas is bubbled through water, the bubbles of gas become saturated with water vapor, thus forming a gaseous mixture. The total pressure in a sample of gas which has been collected by bubbling through water is made up of two components: the pressure of the gas of interest and the pressure of water vapor. The partial pressure of the gas of interest is then the total pressure of the sample minus the vapor pressure of water.

68. The volume of a sample of an ideal gas (at a given temperature and pressure) depends on the total number of moles of gas present, not on the specific identity of the gas. For a gaseous mixture as in this problem, the volume of the gas will depend on the total *combined* moles of nitrogen, oxygen, and helium.

Molar masses: N_2, 28.02 g; O_2, 32.00 g; He, 4.003 g

T = 28 °C = 301 K

$$6.91 \text{ g } N_2 \times \frac{1 \text{ mol}}{28.02 \text{ g}} = 0.2466 \text{ mol } N_2$$

$$4.71 \text{ g } O_2 \times \frac{1 \text{ mol}}{32.00 \text{ g}} = 0.1472 \text{ mol } O_2$$

$$2.95 \text{ g He} \times \frac{1 \text{ mol}}{4.003 \text{ g}} = 0.7369 \text{ mol He}$$

n_{total} = 0.2466 mol + 0.1472 mol + 0.7369 mol = 1.131 mol

$$V = \frac{nRT}{P} = \frac{(1.131 \text{ mol})(0.08206 \text{ L atm mol}^{-1} \text{ K}^{-1})(301 \text{ K})}{(1.05 \text{ atm})} = 26.6 \text{ L}$$

70. Molar masses: O_2, 32.00 g; N_2, 28.02 g

    5.21 kg = 5.21 × 10³ g   4.49 kg = 4.49 × 10³ g

    $$4.49 \times 10^3 \text{ g } O_2 \times \frac{1 \text{ mol}}{32.00 \text{ g}} = 140. \text{ mol } O_2$$

    $$5.21 \times 10^3 \text{ g } N_2 \times \frac{1 \text{ mol}}{28.02 \text{ g}} = 186 \text{ mol } N_2$$

    n_{total} = 140. mol + 186 mol = 326 mol

    T = 24 °C = 297 K

    $$P = \frac{nRT}{V} = \frac{(326 \text{ mol})(0.08206 \text{ L atm mol}^{-1} \text{ K}^{-1})(297 \text{ K})}{(50.0 \text{ L})} = 159 \text{ atm}$$

72. The pressures must be expressed in the same units, either mm Hg or atm.

    $P_{oxygen} = P_{total} - P_{water\ vapor}$

    1.02 atm = 775 mm Hg

    P_{oxygen} = 775 mm Hg − 23.756 mm Hg = 751.244 mm = 751 mm Hg

74. 1.032 atm = 784.3 mm Hg

    $P_{hydrogen}$ = 784.3 mm Hg − 32 mm Hg = 752.3 mm Hg = 0.990 atm

    V = 240 mL = 0.240 L

    T = 30°C + 273 = 303 K

    $$n_{hydrogen} = P_{hydrogen}V/RT = \frac{(0.990 \text{ atm})(0.240 \text{ L})}{(0.08206 \text{ L atm mol}^{-1} \text{ K}^{-1})(303 \text{ K})} = 0.00956 \text{ mol}$$

    $$0.00956 \text{ mol } H_2 \times \frac{1 \text{ mol Zn}}{1 \text{ mol } H_2} = 0.00956 \text{ mol of Zn must have reacted}$$

    molar mass of Zn = 65.38 g

    $$0.00956 \text{ mol Zn} \times \frac{65.38 \text{ g Zn}}{1 \text{ mol Zn}} = 0.625 \text{ g Zn must have reacted}$$

76. A theory is successful if it explains known experimental observations. Theories which have been successful in the past may not be successful in the future (for example, as technology evolves, more sophisticated experiments may be possible in the future).

78. Chemists believe the pressure exerted by a gas sample on the walls of its container arises from collisions between the gas molecules and the walls of the container.

80. no

82. If the temperature of a sample of gas is increased, the average kinetic energy of the particles of gas increases. This means that the speeds of the particles increase. If the particles have a higher speed, they will hit the walls of the container more frequently and with greater force, thereby increasing the pressure.

84. Standard Temperature and Pressure, STP = 0 °C, 1 atm pressure. These conditions were chosen because they are easy to attain and reproduce *experimentally*. The barometric pressure within a laboratory is likely to be near 1 atm most days, and 0°C can be attained with a simple ice bath.

86. Molar mass of P_4 = 123.88 g

$$2.51 \text{ g } P_4 \times \frac{1 \text{ mol}}{123.88 \text{ g}} = 0.02026 \text{ mol } P_4$$

From the balanced chemical equation, the amount of hydrogen needed is

$$0.02026 \text{ mol } P_4 \times \frac{6 \text{ mol } H_2}{1 \text{ mol } P_4} = 0.1216 \text{ mol } H_2$$

T = 25 °C = 298 K P = 753 mm Hg = 0.991 atm

$$V = \frac{nRT}{P} = \frac{(0.1216 \text{ mol})(0.08206 \text{ L atm mol}^{-1} \text{ K}^{-1})(298 \text{ K})}{(0.991 \text{ atm})} = 3.00 \text{ L } H_2$$

88. Molar mass of Mg = 24.31 g

$$1.02 \text{ g Mg} \times \frac{1 \text{ mol}}{24.31 \text{ g}} = 0.0420 \text{ mol Mg}$$

Since the coefficients for Mg and Cl_2 in the balanced equation are the same, for 0.0420 mol of Mg reacting we will need 0.0420 mol of Cl_2

STP: 1.00 atm, 273 K

$$V = \frac{nRT}{P} = \frac{(0.0420 \text{ mol})(0.08206 \text{ L atm mol}^{-1} \text{ K}^{-1})(273 \text{ K})}{(1.00 \text{ atm})} = 0.941 \text{ L } Cl_2$$

90. Molar mass of $CaCO_3$ = 100.08 g

$$4.74 \text{ g } CaCO_3 \times \frac{1 \text{ mol}}{100.08 \text{ g}} = 0.0474 \text{ mol } CaCO_3$$

Since the coefficients of $CaCO_3$ and CO_2 in the balanced chemical equation are the same, when 0.0474 mol $CaCO_3$ is heated, 0.0474 mol CO_2 results.

$T = 26\ °C = 299\ K$

$$V = \frac{nRT}{P} = \frac{(0.0474\ \text{mol})(0.08206\ \text{L atm mol}^{-1}\ \text{K}^{-1})(299\ \text{K})}{(0.997\ \text{atm})} = 1.17\ \text{L CO}_2$$

92. Molar mass of $Mg_3N_2 = 100.95\ g$

$$10.3\ \text{g Mg}_3\text{N}_2 \times \frac{1\ \text{mol}}{100.95\ \text{g}} = 0.102\ \text{mol Mg}_3\text{N}_2$$

From the balanced chemical equation, the amount of NH_3 produced will be

$$0.102\ \text{mol Mg}_3\text{N}_2 \times \frac{2\ \text{mol NH}_3}{1\ \text{mol Mg}_3\text{N}_2} = 0.204\ \text{mol NH}_3$$

$T = 24\ °C = 297\ K$ $P = 752\ \text{mm Hg} = 0.989\ \text{atm}$

$$V = \frac{nRT}{P} = \frac{(0.204\ \text{mol})(0.08206\ \text{L atm mol}^{-1}\ \text{K}^{-1})(297\ \text{K})}{(0.989\ \text{atm})} = 5.03\ \text{L}$$

This assumes that the ammonia was collected dry.

94. Molar masses: O_2, 32.00 g; N_2, 28.02 g

$$26.2\ \text{g O}_2 \times \frac{1\ \text{mol}}{32.00\ \text{g}} = 0.819\ \text{mol O}_2$$

$$35.1\ \text{g N}_2 \times \frac{1\ \text{mol}}{28.02\ \text{g}} = 1.25\ \text{mol N}_2$$

$n_{total} = 0.819\ \text{mol} + 1.25\ \text{mol} = 2.07\ \text{mol}$

$T = 35\ °C = 308\ K$

$P = 755\ \text{mm Hg} = 0.993\ \text{atm}$

$$V = \frac{nRT}{P} = \frac{(2.07\ \text{mol})(0.08206\ \text{L atm mol}^{-1}\ \text{K}^{-1})(308\ \text{K})}{(0.993\ \text{atm})} = 52.7\ \text{L}$$

96. $P_1 = 1.47\ \text{atm}$ $P_2 = 1.00\ \text{atm}$ (Standard Pressure)

    $V_1 = 145\ \text{mL}$ $V_2 = ?\ \text{mL}$

    $T_1 = 44\ °C = 317\ K$ $T_2 = 0\ °C = 273\ K$ (Standard Temperature)

$$V_2 = \frac{P_1 V_1 T_2}{P_2 T_1} = \frac{(1.47\ \text{atm})(145\ \text{mL})(273\ \text{K})}{(1.00\ \text{atm})(317\ \text{K})} = 184\ \text{mL}$$

98. Molar masses: He, 4.003 g; Ne, 20.18 g

$$6.25\ \text{g He} \times \frac{1\ \text{mol}}{4.003\ \text{g}} = 1.561\ \text{mol He}$$

$$4.97\ \text{g Ne} \times \frac{1\ \text{mol}}{20.18\ \text{g}} = 0.2463\ \text{mol Ne}$$

$n_{total} = 1.561\ \text{mol} + 0.2463\ \text{mol} = 1.807\ \text{mol}$

Since 1 mol of an ideal gas occupies 22.4 L at STP, the volume is given by

$$1.807 \text{ mol} \times \frac{22.4 \text{ L}}{1 \text{ mol}} = 40.48 \text{ L} = 40.5 \text{ L}$$

The partial pressure of a given gas in a mixture will be proportional to what *fraction* of the total number of moles of gas the given gas represents

$$P_{He} = \frac{1.561 \text{ mol He}}{1.807 \text{ mol total}} \times 1.00 \text{ atm} = 0.8639 \text{ atm} = 0.864 \text{ atm}$$

$$P_{Ne} = \frac{0.2463 \text{ mol Ne}}{1.807 \text{ mol total}} \times 1.00 \text{ atm} = 0.1363 \text{ atm} = 0.136 \text{ atm}$$

100. Molar mass of N_2 = 28.02 g

$$10.2 \text{ g } N_2 \times \frac{1 \text{ mol}}{28.02 \text{ g}} = 0.364 \text{ mol } N_2$$

$$0.364 \text{ mol } N_2 \times \frac{3 \text{ mol } Cl_2}{1 \text{ mol } N_2} = 1.09 \text{ mol } Cl_2$$

$$1.09 \text{ mol } Cl_2 \times \frac{22.4 \text{ L}}{1 \text{ mol}} = 24.5 \text{ L } Cl_2$$

102. $2K_2MnO_4(aq) + Cl_2(g) \rightarrow 2KMnO_4(s) + 2KCl(aq)$

molar mass $KMnO_4$ = 158.0 g

$$10.0 \text{ g } KMnO_4 \times \frac{1 \text{ mol } KMnO_4}{158.0 \text{ g } KMnO_4} = 0.06329 \text{ mol } KMnO_4$$

$$0.06329 \text{ mol } KMnO_4 \times \frac{1 \text{ mol } Cl_2}{2 \text{ mol } KMnO_4} = 0.03165 \text{ mol } Cl_2$$

$$0.03165 \text{ mol } Cl_2 \times \frac{22.4 \text{ L}}{1 \text{ mol}} = 0.709 \text{ L} = 709 \text{ mL}$$

104. twice

106. a. $PV = k$; $P_1V_1 = P_2V_2$

    b. $V = bT$; $V_1/T_1 = V_2/T_2$

    c. $V = an$; $V_1/n_1 = V_2/n_2$

    d. $PV = nRT$

    e. $P_1V_1/T_1 = P_2V_2/T_2$

164   Chapter 12   Gases

108. First determine what volume the helium in the tank would have if it were at a pressure of 755 mm Hg (corresponding to the pressure the gas will have in the balloons).

8.40 atm = 6384 mm Hg

$$V_2 = (25.2 \text{ L}) \times \frac{6384 \text{ mm Hg}}{755 \text{ mm Hg}} = 213 \text{ L}$$

Allowing for the fact that 25.2 L of He will have to remain in the tank, this leaves 213 - 25.2 = 187.8 L of He for filling the balloons.

$$187.8 \text{ L He} \times \frac{1 \text{ balloon}}{1.50 \text{ L He}} = 125 \text{ balloons}$$

110. According to the balanced chemical equation, when 1 mol of $(NH_4)_2CO_3$ reacts, a total of 4 moles of gaseous substances are produced.

molar mass $(NH_4)_2CO_3$ = 96.09 g

$$52.0 \text{ g} \times \frac{1 \text{ mol}}{96.09 \text{ g}} = 0.541 \text{ mol}$$

Since 0.541 mol of $(NH_4)_2CO_3$ reacts, 4(0.541) = 2.16 mol of gaseous products result.

453 °C = 726 K

$$V = \frac{(2.16 \text{ mol})(0.08206 \text{ L atm/mol}^{-1} \text{ K}^{-1})(726 \text{ K})}{(1.04 \text{ atm})} = 124 \text{ L}$$

112. $CaCO_3(s) + 2H^+(aq) \rightarrow Ca^{2+}(aq) + H_2O(l) + CO_2(g)$

molar mass $CaCO_3$ = 100.1 g

$$10.0 \text{ g CaCO}_3 \times \frac{1 \text{ mol CaCO}_3}{100.1 \text{ g CaCO}_3} = 0.0999 \text{ mol CaCO}_3 = 0.0999 \text{ mol CO}_2 \text{ also}$$

60°C + 273 = 333 K

$P_{\text{carbon dioxide}} = P_{\text{total}} - P_{\text{water vapor}}$

$P_{\text{carbon dioxide}}$ = 774 mm Hg - 149.4 mm Hg = 624.6 mm Hg = 0.822 atm

$$V_{\text{wet}} = \frac{(0.0999 \text{ mol})(0.08206 \text{ L atm mol}^{-1} \text{ K}^{-1})(333 \text{ K})}{(0.822 \text{ atm})} = 3.32 \text{ L wet CO}_2$$

$$V_{\text{dry}} = 3.32 \text{ L} \times \frac{624.6 \text{ mm Hg}}{774 \text{ mm Hg}} = 2.68 \text{ L}$$

114. $2KClO_3(s) \rightarrow 2KCl(s) + 3O_2(g)$

molar mass $KClO_3$ = 122.6 g

$$50.0 \text{ g KClO}_3 \times \frac{1 \text{ mol KClO}_3}{122.6 \text{ g KClO}_3} = 0.408 \text{ mol KClO}_3$$

$$0.408 \text{ mol KClO}_3 \times \frac{3 \text{ mol O}_2}{2 \text{ mol KClO}_3} = 0.612 \text{ mol O}_2$$

$25°C + 273 = 298 \text{ K}$ \qquad 630. torr = 0.829 atm

$$V = nRT/P = \frac{(0.612 \text{ mol})(0.08206 \text{ L atm mol}^{-1} \text{ K}^{-1})(298 \text{ K})}{(0.829 \text{ atm})} = 18.1 \text{ L O}_2$$

116. a. $\quad 752 \text{ mm Hg} \times \dfrac{101,325 \text{ Pa}}{760 \text{ mm Hg}} = 1.00 \times 10^5 \text{ Pa}$

   b. $\quad 458 \text{ kPa} \times \dfrac{1 \text{ atm}}{101.325 \text{ kPa}} = 4.52 \text{ atm}$

   c. $\quad 1.43 \text{ atm} \times \dfrac{760 \text{ mm Hg}}{1 \text{ atm}} = 1.09 \times 10^3 \text{ mm Hg}$

   d. $\quad 842 = 842 \text{ mm Hg}$

118. a. $\quad 645 \text{ mm Hg} \times \dfrac{1 \text{ atm}}{760 \text{ mm Hg}} \times \dfrac{101,325 \text{ Pa}}{1 \text{ atm}} = 8.60 \times 10^4 \text{ Pa}$

   b. $\quad 221 \text{ kPa} = 221 \times 10^3 \text{ Pa} = 2.21 \times 10^5 \text{ Pa}$

   c. $\quad 0.876 \text{ atm} \times \dfrac{101,325 \text{ Pa}}{1 \text{ atm}} = 8.88 \times 10^4 \text{ Pa}$

   d. $\quad 32 \text{ torr} \times \dfrac{1 \text{ atm}}{760 \text{ torr}} \times \dfrac{101,325 \text{ Pa}}{1 \text{ atm}} = 4.3 \times 10^3 \text{ Pa}$

120. a. $\quad 1.00 \text{ mm Hg} = 1.00 \text{ torr}$

   $$V = 255 \text{ mL} \times \frac{1.00 \text{ torr}}{2.00 \text{ torr}} = 128 \text{ mL}$$

   b. $\quad 1.0 \text{ atm} = 101.325 \text{ kPa}$

   $$V = 1.3 \text{ L} \times \frac{1.0 \text{ kPa}}{101.325 \text{ kPa}} = 1.3 \times 10^{-2} \text{ L}$$

   c. $\quad 1.0 \text{ mm Hg} = 0.133 \text{ kPa}$

   $$V = 1.3 \text{ L} \times \frac{1.0 \text{ kPa}}{0.133 \text{ kPa}} = 9.8 \text{ L}$$

122. $1.52 \text{ L} = 1.52 \times 10^3 \text{ mL}$

$$755 \text{ mm Hg} \times \frac{1.52 \times 10^3 \text{ mL}}{450 \text{ mL}} = 2.55 \times 10^3 \text{ mm Hg}$$

124. a. 74°C + 273 = 347 K   −74°C + 273 = 199 K

$$100. \text{ mL} \times \frac{199 \text{ K}}{347 \text{ K}} = 57.3 \text{ mL}$$

b. 100°C + 273 = 373 K

$$373 \text{ K} \times \frac{600 \text{ mL}}{500 \text{ mL}} = 448 \text{ K } (175°C)$$

c. zero (the volume of any gas sample becomes zero at 0 K)

126. 12 °C + 273 = 285 K   192 °C + 273 = 465 K

$$75.2 \text{ mL} \times \frac{465 \text{ K}}{285 \text{ K}} = 123 \text{ mL}$$

128. For a given gas, the number of moles present in a sample is directly proportional to the mass of the sample. So the problem can be solved even though the gas is not identified (so that its molar mass is not known).

$$23.2 \text{ g} \times \frac{10.4 \text{ L}}{93.2 \text{ L}} = 2.59 \text{ g}$$

130. a. $V = 21.2 \text{ mL} = 0.0212 \text{ L}$

$$T = PV/nR = \frac{(1.034 \text{ atm})(0.0212 \text{ L})}{(0.00432 \text{ mol})(0.08206 \text{ L atm mol}^{-1} \text{ K}^{-1})} = 61.8 \text{ K}$$

b. $V = 1.73 \text{ mL} = 0.00173 \text{ L}$

$$P = nRT/V = \frac{(0.000115 \text{ mol})(0.08206 \text{ L atm mol}^{-1} \text{ K}^{-1})(182 \text{ K})}{(0.00173 \text{ L})}$$

$P = 0.993 \text{ atm}$

c. $P = 1.23 \text{ mm Hg} = 0.00162 \text{ atm}$

$T = 152°C + 273 = 425 \text{ K}$

$$V = nRT/P = \frac{(0.773 \text{ mol})(0.08206 \text{ L atm mol}^{-1} \text{ K}^{-1})(425 \text{ K})}{(0.00162 \text{ atm})}$$

$V = 1.66 \times 10^4 \text{ L}$

132. 27°C + 273 = 300 K

The number of moles of gas it takes to fill the 100. L tanks to 120 atm at 27°C is independent of the identity of the gas.

$$n = PV/RT = \frac{(120 \text{ atm})(100. \text{ L})}{(0.08206 \text{ L atm mol}^{-1} \text{ K}^{-1})(300 \text{ K})} = 487 \text{ mol}$$

487 mol of *any* gas will fill the tanks to the required specifications.

molar masses: CH_4, 16.0 g; N_2, 28.0 g; CO_2, 44.0 g

for CH_4: (487 mol)(16.0 g/mol) = 7792 g = 7.79 kg CH_4

for N_2: (487 mol)(28.0 g/mol) = 13,636 g = 13.6 kg N_2

for CO_2: (487 mol)(44.0 g/mol) = 21,428 g = 21.4 kg CO_2

134. molar mass of O_2 = 32.00 g

 55 mg = 0.055 g

 $n = 0.055 \text{ g} \times \dfrac{1 \text{ mol}}{32.00 \text{ g}} = 0.0017 \text{ mol}$

 V = 100. mL = 0.100 L

 T = 26°C + 273 = 299 K

 $P = nRT/V = \dfrac{(0.0017 \text{ mol})(0.08206 \text{ L atm mol}^{-1} \text{ K}^{-1})(299 \text{ K})}{(0.100 \text{ L})} = 0.42 \text{ atm}$

136. P_1 = 1.13 atm P_2 = 1.89 atm

 V_1 = 100 mL = 0.100 L V_2 = 500 mL = 0.500 L

 T_1 = 300 K T_2 = ?

 $T_2 = \dfrac{T_1 P_2 V_2}{P_1 V_1} = \dfrac{(300 \text{ K})(1.89 \text{ atm})(0.500 \text{ L})}{(1.13 \text{ atm})(0.100 \text{ L})} = 2.51 \times 10^3 \text{ K}$

 Note that the calculation could have been carried through with the two volumes expressed in milliliters since the universal gas constant does not appear explicitly in this form of the ideal gas equation.

138. molar masses: N_2, 28.0 g; He, 4.003 g

 $12.1 \text{ g } N_2 \times \dfrac{1 \text{ mol } N_2}{28.0 \text{ g } N_2} = 0.432 \text{ mol } N_2$

 $4.05 \text{ g He} \times \dfrac{1 \text{ mol He}}{4.003 \text{ g He}} = 1.01 \text{ mol He}$

 Total moles of gas = 0.432 mol + 1.01 mol = 1.44 mol

 STP: 1.00 atm, 273 K

 $V = nRT/P = \dfrac{(1.44 \text{ mol})(0.08206 \text{ L atm mol}^{-1} \text{ K}^{-1})(273 \text{ K})}{(1.00 \text{ atm})} = 32.3 \text{ L}$

140. molar mass of NH_3 = 17.03 g

 $5.00 \text{ g } NH_3 \times \dfrac{1 \text{ mol } NH_3}{17.03 \text{ g } NH_3} = 0.294 \text{ mol } NH_3 \text{ to be produced}$

168   Chapter 12   Gases

$$N_2(g) + 3H_2(g) \rightarrow 2NH_3(g)$$

$$0.294 \text{ mol NH}_3 \times \frac{1 \text{ mol N}_2}{2 \text{ mol NH}_3} = 0.147 \text{ mol N}_2 \text{ required}$$

$$0.294 \text{ mol NH}_3 \times \frac{3 \text{ mol H}_2}{2 \text{ mol NH}_3} = 0.441 \text{ mol H}_2 \text{ required}$$

$$11°C + 273 = 284 \text{ K}$$

$$V_{nitrogen} = \frac{(0.147 \text{ mol})(0.08206 \text{ L atm mol}^{-1} \text{ K}^{-1})(284 \text{ K})}{(0.998 \text{ atm})} = 3.43 \text{ L N}_2$$

$$V_{hydrogen} = \frac{(0.441 \text{ mol})(0.08206 \text{ L atm mol}^{-1} \text{ K}^{-1})(284 \text{ K})}{(0.998 \text{ atm})} = 10.3 \text{ L H}_2$$

142. $2Cu_2S(s) + 3O_2(g) \rightarrow 2Cu_2O(s) + 2SO_2(g)$

molar mass $Cu_2S = 159.2$ g

$$25 \text{ g Cu}_2\text{S} \times \frac{1 \text{ mol Cu}_2\text{S}}{159.2 \text{ g Cu}_2\text{S}} = 0.1570 \text{ mol Cu}_2\text{S}$$

$$0.1570 \text{ mol Cu}_2\text{S} \times \frac{3 \text{ mol O}_2}{2 \text{ mol Cu}_2\text{S}} = 0.2355 \text{ mol O}_2$$

$$27.5°C + 273 = 301 \text{ K}$$

$$V_{oxygen} = \frac{(0.2355 \text{ mol})(0.08206 \text{ L atm mol}^{-1} \text{ K}^{-1})(301 \text{ K})}{(0.998 \text{ atm})} = 5.8 \text{ L O}_2$$

$$0.1570 \text{ mol Cu}_2\text{S} \times \frac{2 \text{ mol SO}_2}{2 \text{ mol Cu}_2\text{S}} = 0.1570 \text{ mol SO}_2$$

$$V_{sulfur\ dioxide} = \frac{(0.1570 \text{ mol})(0.08206 \text{ L atm mol}^{-1} \text{ K}^{-1})(301 \text{ K})}{(0.998 \text{ atm})}$$

$$V_{sulfur\ dioxide} = 3.9 \text{ L SO}_2$$

144. One mole of any ideal gas occupies 22.4 L at STP.

$$35 \text{ mol N}_2 \times \frac{22.4 \text{ L}}{1 \text{ mol}} = 7.8 \times 10^2 \text{ L}$$

146. molar masses: He, 4.003 g; Ar, 39.95 g; Ne, 20.18 g

$$5.0 \text{ g He} \times \frac{1 \text{ mol He}}{4.003 \text{ g He}} = 1.249 \text{ mol He}$$

$$1.0 \text{ g Ar} \times \frac{1 \text{ mol Ar}}{39.95 \text{ g Ar}} = 0.02503 \text{ mol Ar}$$

$$3.5 \text{ g Ne} \times \frac{1 \text{ mol Ne}}{20.18 \text{ g Ne}} = 0.1734 \text{ mol Ne}$$

Total moles of gas = 1.249 + 0.02503 + 0.1734 = 1.447 mol

22.4 L is the volume occupied by one mole of any ideal gas at STP. This would apply even if the gas sample is a *mixture* of individual gases.

$$1.447 \text{ mol} \times \frac{22.4 \text{ L}}{1 \text{ mol}} = 32 \text{ L total volume for the mixture}$$

The *partial pressure* of each individual gas in the mixture will be related to what *fraction* on a mole basis each gas represents in the mixture.

$$P_{He} = 1.00 \text{ atm} \times \frac{1.249 \text{ mol He}}{1.447 \text{ mol total}} = 0.86 \text{ atm}$$

$$P_{Ar} = 1.00 \text{ atm} \times \frac{0.02503 \text{ mol Ar}}{1.447 \text{ mol total}} = 0.017 \text{ atm}$$

$$P_{Ne} = 1.00 \text{ atm} \times \frac{0.1734 \text{ mol Ne}}{1.447 \text{ mol total}} = 0.12 \text{ atm}$$

148. The solution is only 50% H_2O_2. Therefore 125 g solution = 62.5 g H_2O_2

    Molar mass of H_2O_2 = 34.02 g

$$62.5 \text{ g } H_2O_2 \times \frac{1 \text{ mol}}{34.02 \text{ g}} = 1.84 \text{ mol } H_2O_2$$

$$1.84 \text{ mol } H_2O_2 \times \frac{1 \text{ mol } O_2}{2 \text{ mol } H_2O_2} = 0.920 \text{ mol } O_2$$

$T = 27 \text{ °C} = 300 \text{ K}$ $\qquad$ $P = 764 \text{ mm Hg} = 1.01 \text{ atm}$

$$V = nRT/P = \frac{(0.920)(0.08206 \text{ L atm mol}^{-1} \text{ K}^{-1})(300 \text{ K})}{(1.01 \text{ atm})} = 22.4 \text{ L}$$

Chapter 13 Liquids and Solids

2. Water exerts its cooling effect in nature in many ways. Water, as perspiration, helps cool the human body (the evaporation of water from skin is an endothermic process; the heat required for evaporation comes from the body). Large bodies of natural water (e.g., the oceans) have a cooling effect on nearby land masses (the interior of the United States, away from the oceans, tends to be hotter than coastal regions). In industry, water is used as a coolant in *many* situations Some nuclear power plants, for example, use water to cool the reactor core. Many office buildings are air-conditioned in summer by circulating cold water systems.

4. The fact that water expands when it freezes often results in broken water pipes during cold weather. The expansion of water when it freezes also makes ice float on liquid water. The expansion of a given mass of water into a larger volume upon freezing lowers the density of ice compared to liquid water. Aquatic life (and probably all life) could not exist if ice sank in water.

6. Sloped portions of a heating/cooling curve represent *changes in temperature* as heat is applied or removed; for example, in the cooling/heating curve shown, there are sloped portions representing the heating of ice, the heating of liquid water, and the heating of steam, as heat continues to be applied. Flat portions of such curves represent equilibrium transitions between states; for example, the flat portions in the curve shown represent the solid-liquid (melting-freezing) transition and the liquid-vapor (boiling-condensation) phase transitions.

8. As a liquid is heated, the motions of the molecules increase as the temperature rises. As the liquid reaches its boiling point, bubbles of vapor begin to form in the liquid, which rise to the surface of the liquid and burst. As the liquid remains at its boiling point, the additional heat energy being supplied to the liquid is used to overcome attractive forces among the molecules in the liquid. As heat energy continues to be applied, more and more molecules will be moving in the right direction and with sufficient energy to escape from the liquid.

10. To melt a solid, or to vaporize a liquid, the molecules of the substance must be moved apart: thus, it is the *inter*molecular forces that must be overcome.

12. It takes more heat to vaporize a liquid than to melt the same amount of solid because of the greater degree to which the intermolecular forces must be overcome.

14. $10.0 \text{ g X} \times \dfrac{1 \text{ mol X}}{52 \text{ g X}} = 0.192 \text{ mol X}$

$0.192 \text{ mol} \times \dfrac{2.5 \text{ kJ}}{1 \text{ mol}} = 0.48 \text{ kJ}$

Chapter 13   Liquids and Solids   171

$$25.0 \text{ g X} \times \frac{1 \text{ mol X}}{52 \text{ g X}} = 0.481 \text{ mol X}$$

$$0.481 \text{ mol} \times \frac{55.3 \text{ kJ}}{1 \text{ mol}} = 27 \text{ kJ}$$

16. molar mass H_2O = 18.02 g

$$25.0 \text{ g } H_2O \times \frac{1 \text{ mol } H_2O}{18.02 \text{ g } H_2O} = 1.39 \text{ mol}$$

To melt the ice: $1.39 \text{ mol} \times \frac{6.02 \text{ kJ}}{1 \text{ mol}} = 8.37 \text{ kJ}$

$$37.5 \text{ g } H_2O \times \frac{1 \text{ mol } H_2O}{18.02 \text{ g } H_2O} = 2.08 \text{ mol}$$

To vaporize the liquid: $2.08 \text{ mol} \times \frac{40.6 \text{ kJ}}{1 \text{ mol}} = 84.4 \text{ kJ}$

To heat the liquid: $55.2 \text{ g} \times 4.18 \frac{J}{g \, °C} \times 100°C = 23{,}073 \text{ J} = 23.1 \text{ kJ}$

18. The *molar* heat of fusion of aluminum is the heat required to melt 1 mol.

$$\frac{3.95 \text{ kJ}}{1 \text{ g}} \times \frac{26.98 \text{ g}}{1 \text{ mol}} = 107 \text{ kJ/mol}$$

$$10.0 \text{ g} \times \frac{3.95 \text{ kJ}}{1 \text{ g}} = 39.5 \text{ kJ required to melt } 10.0 \text{ g Al}$$

$$10.0 \text{ mol} \times \frac{107 \text{ kJ}}{1 \text{ mol}} = 1.07 \times 10^3 \text{ kJ required to melt } 10.0 \text{ mol Al.}$$

20. Dipole-dipole forces are relatively stronger at short distances; they are short-range forces. Molecules must first closely approach one another before dipole-dipole forces can cause attraction between molecules.

22. Water molecules are able to form strong *hydrogen bonds* with each other. These bonds are an especially strong form of dipole-dipole forces and are only possible when hydrogen atoms are bonded to the most electronegative elements (N, O, and F). The extra strong intermolecular forces in H_2O require much higher temperatures (high energies) to be overcome in order to permit the liquid to boil. We take the fact that water has a much higher boiling point than the other hydrogen compounds of the Group 6 elements as proof that a special force is at play in water (hydrogen bonding).

172   Chapter 13   *Liquids and Solids*

24. The fact that such nonpolar, monatomic atoms *can* be liquefied and solidified indicates that there must be *some* sort of intermolecular forces possible between atoms in these substances. London dispersion forces arise when a temporary (instantaneous) dipolar arrangement of charge develops as the electrons of an atom move around its nucleus. This instantaneous dipole can induce a similar dipole in a neighboring atom, leading to a momentary attraction.

26. a.   London dispersion forces (nonpolar, atoms)

    b.   London dispersion forces (nonpolar molecules)

    c.   dipole-dipole forces; London dispersion forces

    d.   hydrogen bonding (H attached to O); London dispersion forces

28. An increase in the heat of fusion is observed for an increase in the size of the halogen atom involved (the electron cloud of a larger atom is more easily polarized by an approaching dipole, thus giving larger London dispersion forces).

30. For a homogeneous mixture to be able to form at all, the forces between molecules of the two substances being mixed must be at least *comparable in magnitude* to the intermolecular forces within each *separate* substance. Apparently in the case of a water-ethanol mixture, the forces that exist when water and ethanol are mixed are stronger than water-water or ethanol-ethanol forces in the separate substances. This allows ethanol and water molecules to approach each other more closely in the mixture than either substance's molecules could approach a like molecule in the separate substances. There is strong hydrogen bonding in both ethanol and water.

32. Vapor pressure is the pressure of vapor present *at equilibrium* above a liquid in a sealed container at a particular temperature. When a liquid is placed in a closed container, molecules of the liquid evaporate freely into the empty space above the liquid. As the number of molecules present in the vapor state increases with time, vapor molecules begin to rejoin the liquid state (condense). Eventually a dynamic equilibrium is reached between evaporation and condensation in which the net number of molecules present in the vapor phase becomes *constant* with time.

34. A method is shown in Figure 13.10. The apparatus consists of basically a barometer into which a volatile liquid may be injected. Since mercury is so much more dense than other liquids, the injected volatile liquid rises to the top of the mercury column in the tube and floats on top of the mercury. Since the space above the mercury is a vacuum, the volatile liquid evaporates into the empty space. As the liquid is converted to the gaseous state, the level of the mercury column drops as the pressure of vapor builds up.

36. a.   HF   Although both substances are capable of hydrogen bonding, water has two O-H bonds which can be involved in hydrogen bonding versus only one F-H bond in HF.

Chapter 13   Liquids and Solids   173

    b.   CH$_3$OCH$_3$ Since there is no H attached to the O atom, no hydrogen bonding can exist. Since there is no hydrogen bonding possible, the molecule should be relatively more volatile than CH$_3$CH$_2$OH even though it contains the same number of atoms of each element.

    c.   CH$_3$SH   Hydrogen bonding is not as important for a S-H bond (because S has a lower electronegativity than O). Since there is little hydrogen bonding, the molecule is relatively more volatile than CH$_3$OH.

38. Both substances have the same molar mass. However ethyl alcohol contains a hydrogen atom directly bonded to an oxygen atom. Therefore, hydrogen bonding can exist in ethyl alcohol, whereas only weak dipole-dipole forces can exist in dimethyl ether. Dimethyl ether is more volatile; ethyl alcohol has a higher boiling point.

40. *Ionic* solids have as their fundamental particles positive and negative ions; a simple example is sodium chloride, in which Na$^+$ and Cl$^-$ ions are held together by strong electrostatic forces.

Molecular solids have molecules as their fundamental particles, with the molecules being held together in the crystal by dipole-dipole forces, hydrogen bonding forces, or London dispersion forces (depending on the identity of the substance); simple examples of molecular solids include ice (H$_2$O) and ordinary table sugar (sucrose).

Atomic solids have simple atoms as their fundamental particles, with the atoms being held together in the crystal either by covalent bonding (as in graphite or diamond) or by metallic bonding (as in copper or other metals).

42. The interparticle forces in ionic solids (the ionic bond) are much stronger than the interparticle forces in molecular solids (dipole-dipole forces, London forces, etc.). The difference in intermolecular forces is most clearly shown in the great differences in melting points and boiling points between ionic and molecular solids. For example, table salt and ordinary sugar are both crystalline solids that appear very similar. Yet sugar can be melted easily in a saucepan during the making of candy, whereas even the full heat of a stove will not melt salt.

44. Ionic solids consist of a crystal lattice of basically alternating positively and negatively charged ions. A given ion is surrounded by several ions of the opposite charge, all of which electrostatically attract it strongly. This pattern repeats itself throughout the crystal. The existence of these strong electrostatic forces throughout the crystal means a great deal of energy my be applied to overcome the forces and melt the solid.

174   Chapter 13   Liquids and Solids

46. Ordinary ice contains nonlinear, highly polar water molecules. In addition, there is extensive, strong hydrogen bonding possible between water molecules in ordinary ice. Dry ice, on the other hand, consists of linear, nonpolar carbon dioxide molecules, and only very weak intermolecular forces are possible.

48. In a metal, the valence electrons are mobile, and can move throughout the entire metal's crystal lattice. In ionic solids, although there are positive and negative ions present, each ion is held rigidly in place by several ions of the opposite charge.

50. Alloys may be of two types: *substitutional* (in which one metal is substituted for another in the regular positions of the crystal lattice) and *interstitial* (in which a second metal's atoms fit into the empty space in a given metal's crystal lattice). The presence of atoms of a second metal in a given metal's crystal lattice changes the properties of the metal: frequently the alloy is stronger than either of the original metals because the irregularities introduced into the crystal lattice by the presence of a second metal's atoms prevent the crystal from being deformed as easily. The properties of iron may be modified by alloying with many different substances, particularly with carbon, nickel, and cobalt. Steels with relatively high carbon content are exceptionally strong, whereas steels with low carbon contents are softer, more malleable, and more ductile. Steels produced by alloying iron with nickel and cobalt are more resistant to corrosion than iron itself.

52. j

54. f

56. d

58. a

60. l

62. Dimethyl ether has the larger vapor pressure. No hydrogen bonding is possible since the O atom does not have a hydrogen atom attached. Hydrogen bonding can occur *only* when a hydrogen atom is *directly* attached to a strongly electronegative atom (such as N, O, or F). Hydrogen bonding *is* possible in ethanol (ethanol contains an -OH group).

64. a. H_2. London dispersion forces are the only intermolecular forces present in these nonpolar molecules; typically London forces become larger with increasing atomic size (as the atoms become bigger, the edge of the electron cloud lies farther from the nucleus and becomes more easily distorted).

　　 b. Xe. Only the relatively weak London forces could exist in a crystal of Xe atoms, whereas in NaCl strong ionic forces exist, and in diamond strong covalent bonding exists between carbon atoms.

　　 c. Cl_2. Only London forces exist among such nonpolar molecules. London forces become larger with increasing atomic size.

66. Steel is a general term applied to alloys consisting primarily of iron, but with small amounts of other substances added. Whereas pure iron itself is relatively soft, malleable, and ductile, steels are typically much stronger and harder, and much less subject to damage.

68. Water is the solvent in which cellular processes take place in living creatures. Water in the oceans moderates the earth's temperature. Water is used in industry as a cooling agent. Water serves as a means of transportation on the earth's oceans. The liquid range is 0°C to 100°C at 1 atm pressure.

70. At higher altitudes, the boiling points of liquids, such as water, are lower because there is a lower atmospheric pressure above the liquid. The temperature at which food cooks is determined by the temperature to which the water in the food can be heated before it escapes as steam. Thus, food cooks at a lower temperature at high elevations where the boiling point of water is lowered.

72. Heat of fusion (melt); Heat of vaporization (boil).
The heat of vaporization is always larger, because virtually all of the intermolecular forces must be overcome to form a gas. In a liquid, considerable intermolecular forces remain. Thus going from a solid to liquid requires less energy than going from the liquid to the gas.

74. Dipole-dipole interactions are typically about 1% as strong as a covalent bond. Dipole-dipole interactions represent electrostatic attractions between portions of molecules which carry only a *partial* positive or negative charge, and such forces require the molecules that are interacting to come *near* enough to each other.

76. London dispersion forces are relatively weak forces that arise among noble gas atoms and in nonpolar molecules. London forces are due to *instantaneous dipoles* that develop when one atom (or molecule) momentarily distorts the electron cloud of another atom (or molecule). London forces are typically weaker than either permanent dipole-dipole forces or covalent bonds.

78. For every mole of liquid water that evaporates, several kilojoules of heat must be absorbed to provide kinetic energy to overcome attractive forces among the molecules. This heat is absorbed by the water from its surroundings.

80. In NH_3, strong hydrogen bonding can exist. In CH_4, because the molecule is nonpolar, only the relatively weak London dispersion forces exist.

82. In a crystal of ice, strong *hydrogen bonding* forces are present, while in the crystal of a nonpolar substance like oxygen, only the much weaker *London* forces exist.

Chapter 14   Solutions

2. A heterogeneous mixture does not have a uniform composition: the composition varies in different places within the mixture. Examples of nonhomogeneous mixtures include salad dressing (mixture of oil, vinegar, water, herbs and spices) and granite (combination of minerals).

4. solid

6. One substance will mix with and dissolve in another substance if the intermolecular forces are similar in the two substances, so that when the mixture forms, the forces between particles in the mixture will be similar to the forces present in the separate substances. Sugar and ethyl alcohol molecules both contain polar -OH groups, which are comparable to the polar -OH structure in water. Sugar or ethyl alcohol molecules can hydrogen bond with water molecules and intermingle with them freely to form a solution. Substances like petroleum (whose molecules contain only carbon and hydrogen) are very nonpolar and cannot form interactions with polar water molecules.

8. independently

10. unsaturated

12. large

14. 100.0

16. a. $\dfrac{5.00 \text{ g CaCl}_2}{(95.0 \text{ g H}_2\text{O} + 5.00 \text{ g CaCl}_2)} \times 100 = 5.00\% \text{ CaCl}_2$

   b. $\dfrac{1.00 \text{ g CaCl}_2}{(19.0 \text{ g H}_2\text{O} + 1.00 \text{ g CaCl}_2)} \times 100 = 5.00\% \text{ CaCl}_2$

   c. $\dfrac{15.0 \text{ g CaCl}_2}{(285 \text{ g H}_2\text{O}) + 15.0 \text{ g CaCl}_2)} \times 100 = 5.00\% \text{ CaCl}_2$

   d. $\dfrac{0.00200 \text{ g CaCl}_2}{(0.0380 \text{ g H}_2\text{O} + 0.00200 \text{ g CaCl}_2)} \times 100 = 5.00\% \text{ CaCl}_2$

18. To say a solution is 15.0% NaCl by mass means that 100. g of the solution would contain 15.0 g of NaCl.

   a. $150. \text{ g solution} \times \dfrac{15.0 \text{ g NaCl}}{100. \text{ g solution}} = 22.5 \text{ g NaCl}$

   b. $35.0 \text{ g NaCl} \times \dfrac{100. \text{ g solution}}{15.0 \text{ g NaCl}} = 233 \text{ g solution}$

c. $\quad 1000.\text{ g solution} \times \dfrac{15.0 \text{ g NaCl}}{100.\text{ g solution}} = 150.\text{ g NaCl}$

d. $\quad 1000.\text{ g solution} \times \dfrac{15.0 \text{ g NaCl}}{100.\text{ g solution}} = 150.\text{ g NaCl}$

Note that the calculation for parts (c) and (d) are identical. These questions are just different ways of asking the same thing: How many g of NaCl are in 1000. g of a 15.0% NaCl solution?

20. $\%\text{Cu} = \dfrac{5.31 \text{ g Cu}}{(5.31 \text{ g Cu} + 4.03 \text{ g Zn} + 145 \text{ g Fe})} \times 100 = \dfrac{5.31 \text{ g}}{154.3 \text{ g}} \times 100 = 3.44\% \text{ Cu}$

$\%\text{ Zn} = \dfrac{4.03 \text{ g Zn}}{(5.31 \text{ g Cu} + 4.03 \text{ g Zn} + 145 \text{ g Fe})} \times 100 = \dfrac{4.03 \text{ g}}{154.3 \text{ g}} \times 100 = 2.61\% \text{ Zn}$

$\%\text{ Fe} = \dfrac{145 \text{ g Fe}}{(5.31 \text{ g Cu} + 4.03 \text{ g Zn} + 145 \text{ g Fe})} \times 100 = \dfrac{145 \text{ g}}{154.3 \text{ g}} \times 100 = 94.0\% \text{ Fe}$

22. $\dfrac{67.1 \text{ g CaCl}_2}{(67.1 \text{ g CaCl}_2 + 275 \text{ g H}_2\text{O})} \times 100 = 19.6\% \text{ CaCl}_2$

24. To say that the solution is 6.25% KBr by mass, means that 100. g of the solution will contain 6.25 g KBr.

$125 \text{ g solution} \times \dfrac{6.25 \text{ g KBr}}{100.\text{ g solution}} = 7.81 \text{ g KBr}$

26. To say that the solution is to be 1.25% $CuCl_2$ by mass, means that 100. g of the solution will contain 1.25 g $CuCl_2$

$1250.\text{ g solution} \times \dfrac{1.25 \text{ g CuCl}_2}{100.\text{ g solution}} = 15.6 \text{ g CuCl}_2$

28. $\text{g heptane} = 93 \text{ g solution} \times \dfrac{5.2 \text{ g heptane}}{100.\text{ g solution}} = 4.8 \text{ g heptane}$

$\text{g pentane} = 93 \text{ g solution} \times \dfrac{2.9 \text{ g pentane}}{100.\text{ g solution}} = 2.7 \text{ g pentane}$

g hexane = 93 g solution − 4.8 g heptane − 2.7 g pentane = 86 g hexane

30. 0.221 mol Ca^{2+}; 0.442 mol Cl^-

32. To say that a solution has a concentration of 5 *M* means that in 1 L of solution (*not* solvent) there would be 5 mol of solute: to prepare such a solution one would place 5 mol of NaCl in a 1 L flask, and then add whatever amount of water is necessary so that the *total* volume would be 1 L after mixing. The NaCl will occupy some space, so the amount of water to be added will be *less* than 1.00 L.

178   Chapter 14   Solutions

34. Molarity = $\dfrac{\text{moles of solute}}{\text{liters of solution}}$

   a. 250 mL = 0.25 L

   $M = \dfrac{0.50 \text{ mol KBr}}{0.25 \text{ L solution}} = 2.0\ M$

   b. 500 mL = 0.500 L

   $M = \dfrac{0.50 \text{ mol KBr}}{0.500 \text{ L solution}} = 1.0\ M$

   c. 750 mL = 0.75 L

   $M = \dfrac{0.50 \text{ mol KBr}}{0.75 \text{ L solution}} = 0.67\ M$

   d. $M = \dfrac{0.50 \text{ mol KBr}}{1.0 \text{ L solution}} = 0.50\ M$

36. Molarity = $\dfrac{\text{moles of solute}}{\text{liters of solution}}$

   a. Molar mass of $CuCl_2$ = 134.45 g   125 mL = 0.125 L

   $4.25 \text{ g } CuCl_2 \times \dfrac{1 \text{ mol}}{134.45 \text{ g}} = 0.0316 \text{ mol } CuCl_2$

   $M = \dfrac{0.0316 \text{ mol } CuCl_2}{0.125 \text{ L solution}} = 0.253\ M$

   b. Molar mass of $NaHCO_3$ = 84.01 g   11.3 mL = 0.0113 L

   $0.101 \text{ g } NaHCO_3 \times \dfrac{1 \text{ mol}}{84.01 \text{ g}} = 0.00120 \text{ mol } NaHCO_3$

   $M = \dfrac{0.00120 \text{ mol } NaHCO_3}{0.0113 \text{ L solution}} = 0.106\ M$

   c. Molar mass of Na_2CO_3 = 105.99 g

   $52.9 \text{ g } Na_2CO_3 \times \dfrac{1 \text{ mol}}{105.99 \text{ g}} = 0.499 \text{ mol } Na_2CO_3$

   $M = \dfrac{0.499 \text{ mol } Na_2CO_3}{1.15 \text{ L solution}} = 0.434\ M$

   d. Molar mass of KOH = 56.11 g   1.5 mL = 0.0015 L

   $0.14 \text{ mg KOH} \times \dfrac{1 \text{ g}}{10^3 \text{ mg}} \times \dfrac{1 \text{ mol}}{56.11 \text{ g}} = 2.50 \times 10^{-6} \text{ mol KOH}$

$$M = \frac{2.50 \times 10^{-6} \text{ mol KOH}}{0.0015 \text{ L solution}} = 1.67 \times 10^{-3} \, M = 1.7 \times 10^{-3} \, M$$

38. Molar mass of $CaBr_2$ = 199.9 g

    $$4.25 \text{ g CaBr}_2 \times \frac{1 \text{ mol}}{199.9 \text{ g}} = 0.0213 \text{ mol CaBr}_2$$

    125 mL = 0.125 L

    $$M = \frac{0.0213 \text{ mol CaBr}_2}{0.125 \text{ L solution}} = 0.170 \, M$$

40. Molar mass of I_2 = 253.8 g

    225 mL = 0.225 L

    $$5.15 \text{ g I}_2 \times \frac{1 \text{ mol}}{253.8 \text{ g}} = 0.0203 \text{ mol I}_2$$

    $$M = \frac{0.0203 \text{ mol I}_2}{0.225 \text{ L solution}} = 0.0902 \, M$$

42. a. Molar mass of NaOH = 40.00 g

    $$495 \text{ g NaOH} \times \frac{1 \text{ mol}}{40.00 \text{ g}} = 12.4 \text{ mol NaOH}$$

    $$M = \frac{12.4 \text{ mol NaOH}}{20.0 \text{ L solution}} = 0.619 \, M$$

44. a. molar mass of HNO_3 = 63.02 g   127 mL = 0.127 L

    $$0.127 \text{ L solution} \times \frac{0.105 \text{ mol HNO}_3}{1.00 \text{ L solution}} = 0.0133 \text{ mol HNO}_3$$

    $$0.0133 \text{ mol HNO}_3 \times \frac{63.02 \text{ g HNO}_3}{1 \text{ mol HNO}_3} = 0.838 \text{ g HNO}_3$$

    b. molar mass of NH_3 = 17.03 g   155 mL = 0.155 L

    $$0.155 \text{ L solution} \times \frac{15.1 \text{ mol NH}_3}{1.00 \text{ L solution}} = 2.34 \text{ mol NH}_3$$

    $$2.34 \text{ mol NH}_3 \times \frac{17.03 \text{ g NH}_3}{1 \text{ mol NH}_3} = 39.9 \text{ g NH}_3$$

    c. molar mass KSCN = 97.19 g

    $$2.51 \text{ L solution} \times \frac{2.01 \times 10^{-3} \text{ mol KSCN}}{1.00 \text{ L solution}} = 5.05 \times 10^{-3} \text{ mol KSCN}$$

    $$5.05 \times 10^{-3} \text{ mol KSCN} \times \frac{97.19 \text{ g KSCN}}{1 \text{ mol KSCN}} = 0.490 \text{ g KSCN}$$

d. molar mass of HCl = 36.46 g 12.2 mL = 0.0122 L

$$0.0122 \text{ L solution} \times \frac{2.45 \text{ mol HCl}}{1.00 \text{ L solution}} = 0.0299 \text{ mol HCl}$$

$$0.0299 \text{ mol HCl} \times \frac{36.46 \text{ g HCl}}{1 \text{ mol HCl}} = 1.09 \text{ g HCl}$$

46. a. molar mass of KBr = 119.0 g 173 mL = 0.173 L

$$0.173 \text{ L solution} \times \frac{1.24 \text{ mol KBr}}{1.00 \text{ L solution}} = 0.215 \text{ mol KBr}$$

$$0.215 \text{ mol KBr} \times \frac{119.0 \text{ g}}{1 \text{ mol}} = 25.6 \text{ g KBr}$$

b. molar mass of HCl = 36.46 g

$$2.04 \text{ L solution} \times \frac{12.1 \text{ mol HCl}}{1.00 \text{ L solution}} = 24.7 \text{ mol HCl}$$

$$24.7 \text{ mol HCl} \times \frac{36.46 \text{ g}}{1 \text{ mol}} = 901 \text{ g HCl}$$

c. molar mass of NH_3 = 17.03 g 25 mL = 0.025 L

$$0.025 \text{ L solution} \times \frac{3.0 \text{ mol } NH_3}{1.00 \text{ L solution}} = 0.075 \text{ mol } NH_3$$

$$0.075 \text{ mol } NH_3 \times \frac{17.03 \text{ g}}{1 \text{ mol}} = 1.3 \text{ g } NH_3$$

d. molar mass $CaCl_2$ = 111.0 g 125 mL = 0.125 L

$$0.125 \text{ L solution} \times \frac{0.552 \text{ mol } CaCl_2}{1.00 \text{ L solution}} = 0.0690 \text{ mol } CaCl_2$$

$$0.0690 \text{ mol } CaCl_2 \times \frac{111.0 \text{ g}}{1 \text{ mol}} = 7.66 \text{ g } CaCl_2$$

48. Molar mass of NaCl = 58.44 g 1 lb = 453.59 g

$$453.59 \text{ g NaCl} \times \frac{1 \text{ mol}}{58.44 \text{ g}} = 7.76 \text{ mol NaCl}$$

$$7.76 \text{ mol NaCl} \times \frac{1.00 \text{ L solution}}{1.0 \text{ mol NaCl}} = 7.76 \text{ L} = 7.8 \text{ L of solution}$$

50. a. 10.2 mL = 0.0102 L

$$0.0102 \text{ L} \times \frac{0.451 \text{ mol } AlCl_3}{1.00 \text{ L}} \times \frac{1 \text{ mol } Al^{3+}}{1 \text{ mol } AlCl_3} = 4.60 \times 10^{-3} \text{ mol } Al^{3+}$$

$$0.0102 \text{ L} \times \frac{0.451 \text{ mol AlCl}_3}{1.00 \text{ L}} \times \frac{3 \text{ mol Cl}^-}{1 \text{ mol AlCl}_3} = 1.38 \times 10^{-2} \text{ mol Cl}^-$$

b. $$5.51 \text{ L} \times \frac{0.103 \text{ mol Na}_3\text{PO}_4}{1.00 \text{ L}} \times \frac{3 \text{ mol Na}^+}{1 \text{ mol Na}_3\text{PO}_4} = 1.70 \text{ mol Na}^+$$

$$5.51 \text{ L} \times \frac{0.103 \text{ mol Na}_3\text{PO}_4}{1.00 \text{ L}} \times \frac{1 \text{ mol PO}_4^{3-}}{1 \text{ mol Na}_3\text{PO}_4} = 0.568 \text{ mol PO}_4^{3-}$$

c. 1.75 mL = 0.00175 L

$$0.00175 \text{ L} \times \frac{1.25 \text{ mol CuCl}_2}{1.00 \text{ L}} \times \frac{1 \text{ mol Cu}^{2+}}{1 \text{ mol CuCl}_2} = 2.19 \times 10^{-3} \text{ mol Cu}^{2+}$$

$$0.00175 \text{ L} \times \frac{1.25 \text{ mol CuCl}_2}{1.00 \text{ L}} \times \frac{2 \text{ mol Cl}^-}{1 \text{ mol CuCl}_2} = 4.38 \times 10^{-3} \text{ mol Cl}^-$$

d. 25.2 mL = 0.0252 L

$$0.0252 \text{ L} \times \frac{0.00157 \text{ mol Ca(OH)}_2}{1.00 \text{ L}} \times \frac{1 \text{ mol Ca}^{2+}}{1 \text{ mol Ca(OH)}_2} = 3.96 \times 10^{-5} \text{ mol Ca}^{2+}$$

$$0.0252 \text{ L} \times \frac{0.00157 \text{ mol Ca(OH)}_2}{1.00 \text{ L}} \times \frac{2 \text{ mol OH}^-}{1 \text{ mol Ca(OH)}_2} = 7.91 \times 10^{-5} \text{ mol OH}^-$$

52. Molar mass of Na_2CO_3 = 106.0 g   250 mL = 0.250 L

$$0.250 \text{ L} \times \frac{0.0500 \text{ mol Na}_2\text{CO}_3}{1.00 \text{ L}} \times \frac{106.0 \text{ g}}{1 \text{ mol}} = 1.33 \text{ g Na}_2\text{CO}_3$$

54. half

56. $M_1 \times V_1 = M_2 \times V_2$

    a. $M_1 = 0.251\ M$ $M_2 = ?$

        $V_1 = 125$ mL $V_2 = 250. + 125 = 375$ mL

$$M_2 = \frac{(0.251\ M)(125 \text{ mL})}{(375 \text{ mL})} = 0.0837\ M$$

    b. $M_1 = 0.499\ M$ $M_2 = ?$

        $V_1 = 445$ mL $V_2 = 445 + 250. = 695$ mL

$$M_2 = \frac{(0.499\ M)(445 \text{ mL})}{(695 \text{ mL})} = 0.320\ M$$

    c. $M_1 = 0.101\ M$ $M_2 = ?$

        $V_1 = 5.25$ L $V_2 = 5.25 + 0.250 = 5.50$ L

$$M_2 = \frac{(0.101\ M)(5.25 \text{ L})}{(5.50 \text{ L})} = 0.0964\ M$$

182 Chapter 14 Solutions

d. $M_1 = 14.5\ M$ $M_2 = ?$
    $V_1 = 11.2$ mL $V_2 = 11.2 + 250. = 261.2$ mL

$$M_2 = \frac{(14.5\ M)(11.2\ \text{mL})}{(261.2\ \text{mL})} = 0.622\ M$$

58. $M_1 = 18.1\ M$ $M_2 = 0.100\ M$
    $V_1 = ?$ mL $V_2 = 125$ mL

$$M_1 = \frac{(0.100\ M)(125\ \text{mL})}{(18.1\ M)} = 0.691\ \text{mL}$$

60. $M_1 = 0.211\ M$ $M_2 = ?$
    $V_1 = 75$ mL $V_2 = 125$ mL

$$M_2 = \frac{(0.211\ M)(75\ \text{mL})}{(125\ \text{mL})} = 0.127\ M = 0.13\ M$$

62. $M_1 = 0.227\ M$ $M_2 = 0.105\ M$
    $V_1 = 25.0$ mL $V_2 = ?$

$$V_2 = \frac{(0.227\ M)(25.0\ \text{mL})}{(0.105\ M)} = 54.0\ \text{mL}$$

So 54.0 − 25.0 = 29.0 mL of water should be added.

64. $Ba(NO_3)_2(aq) + Na_2SO_4(aq) \rightarrow BaSO_4(s) + 2NaNO_3(aq)$

    12.5 mL = 0.0125 L

$$\text{moles } Ba(NO_3)_2 = 0.0125\ \text{L} \times \frac{0.15\ \text{mol } Ba(NO_3)_2}{1.00\ \text{L}} = 1.88 \times 10^{-3}\ \text{mol } Ba(NO_3)_2$$

From the balanced chemical equation for the reaction, if 1.88×10^{-3} mol $Ba(NO_3)_2$ are to be precipitated, then 1.88×10^{-3} mol Na_2SO_4 will be needed.

$$1.88 \times 10^{-3}\ \text{mol } Na_2SO_4 \times \frac{1.00\ \text{L}}{0.25\ \text{mol } Na_2SO_4} = 0.0075\ \text{L required} = 7.5\ \text{mL}$$

66. Molar mass $Na_2C_2O_4 = 134.0$ g   37.5 mL = 0.0375 L

$$\text{moles } Ca^{2+}\ \text{ion} = 0.0375\ \text{L} \times \frac{0.104\ \text{mol}}{1.00\ \text{L}} = 0.00390\ \text{mol } Ca^{2+}\ \text{ion}$$

$Ca^{2+}(aq) + C_2O_4^{2-}(aq) \rightarrow CaC_2O_4(s)$

Since the precipitation reaction is of 1:1 stoichiometry, then 0.00390 mol of $C_2O_4^{2-}$ ion is needed. And since each formula unit of $Na_2C_2O_4$ contains one $C_2O_4^{2-}$ ion, then 0.00390 mol of $Na_2C_2O_4$ is required.

$$0.00390\ \text{mol } Na_2C_2O_4 \times \frac{134.0\ \text{g}}{1\ \text{mol}} = 0.523\ \text{g } Na_2C_2O_4\ \text{required}$$

68. 10.0 mL = 0.0100 L

$$0.0100 \text{ L} \times \frac{0.250 \text{ mol AlCl}_3}{1.00 \text{ L}} = 2.50 \times 10^{-3} \text{ mol AlCl}_3$$

$\text{AlCl}_3(aq) + 3\text{NaOH}(s) \rightarrow \text{Al(OH)}_3(s) + 3\text{NaCl}(aq)$

$$2.50 \times 10^{-3} \text{ mol AlCl}_3 \times \frac{3 \text{ mol NaOH}}{1 \text{ mol AlCl}_3} = 7.50 \times 10^{-3} \text{ mol NaOH}$$

molar mass NaOH = 40.0 g

$$7.50 \times 10^{-3} \text{ mol NaOH} \times \frac{40.0 \text{ g NaOH}}{1 \text{ mol}} = 0.300 \text{ g NaOH}$$

70. $\text{HCl}(aq) + \text{NaOH}(aq) \rightarrow \text{NaCl}(aq) + \text{H}_2\text{O}(l)$

24.9 mL = 0.0249 L

$$0.0249 \text{ L} \times \frac{0.451 \text{ mol NaOH}}{1.00 \text{ L}} = 0.0112 \text{ mol NaOH}$$

$$0.0112 \text{ mol NaOH} \times \frac{1 \text{ mol HCl}}{1 \text{ mol NaOH}} = 0.0112 \text{ mol HCl}$$

$$0.0112 \text{ mol HCl} \times \frac{1.00 \text{ L}}{0.175 \text{ mol HCl}} = 0.0642 \text{ L required} = 64.2 \text{ mL}$$

72. 7.2 mL = 0.0072 L

$$0.0072 \text{ L} \times \frac{2.5 \times 10^{-3} \text{ mol NaOH}}{1.00 \text{ L}} = 1.8 \times 10^{-5} \text{ mol NaOH}$$

$\text{H}^+(aq) + \text{OH}^-(aq) \rightarrow \text{H}_2\text{O}(l)$

$$1.8 \times 10^{-5} \text{ mol OH}^- \times \frac{1 \text{ mol H}^+}{1 \text{ mol OH}^-} = 1.8 \times 10^{-5} \text{ mol H}^+$$

100 mL = 0.100 L

$$M = \frac{1.8 \times 10^{-5} \text{ mol H}^+}{0.100 \text{ L}} = 1.8 \times 10^{-4} \; M \; \text{H}^+(aq)$$

74. Experimentally, neutralization reactions are usually performed with volumetric glassware that is calibrated in milliliters rather than liters. For convenience in calculations for such reactions, the arithmetic is often performed in terms of *milli*liters and *milli*moles, rather than in liters and moles. 1 mmol = 0.001 mol. Note that the number of moles of solute per liter of solution, the molarity, is numerically equivalent to the number of *milli*moles of solute per *milli*liter of solution.

 a. $\text{HNO}_3(aq) + \text{NaOH}(aq) \rightarrow \text{NaNO}_3(aq) + \text{H}_2\text{O}(l)$

$$12.7 \text{ mL} \times \frac{0.501 \text{ mmol}}{1 \text{ mL}} = 6.36 \text{ mmol NaOH present in the sample}$$

184   Chapter 14   Solutions

$$6.36 \text{ mmol NaOH} \times \frac{1 \text{ mmol HNO}_3}{1 \text{ mmol NaOH}} = 6.36 \text{ mmol HNO}_3 \text{ required to react}$$

$$6.36 \text{ mmol HNO}_3 \times \frac{1.00 \text{ mL}}{0.101 \text{ mmol HNO}_3} = 63.0 \text{ mL HNO}_3 \text{ required}$$

b. $2HNO_3(aq) + Ba(OH)_2 \rightarrow Ba(NO_3)_2 + 2H_2O(l)$

$$24.9 \text{ mL} \times \frac{0.00491 \text{ mmol}}{1.00 \text{ mL}} = 0.122 \text{ mmol Ba(OH)}_2 \text{ present in the sample}$$

$$0.122 \text{ mmol Ba(OH)}_2 \times \frac{2 \text{ mmol HNO}_3}{1 \text{ mmol Ba(OH)}_2} = 0.244 \text{ mmol HNO}_3 \text{ required}$$

$$0.244 \text{ mmol HNO}_3 \times \frac{1.00 \text{ mL}}{0.101 \text{ mmol HNO}_3} = 2.42 \text{ mL HNO}_3 \text{ is required}$$

c. $HNO_3(aq) + NH_3(aq) \rightarrow NH_4NO_3(aq)$

$$49.1 \text{ mL} \times \frac{0.103 \text{ mmol}}{1.00 \text{ mL}} = 5.06 \text{ mmol NH}_3 \text{ present in the sample}$$

$$5.06 \text{ mmol NH}_3 \times \frac{1 \text{ mmol HNO}_3}{1 \text{ mmol NH}_3} = 5.06 \text{ mmol HNO}_3 \text{ required}$$

$$5.06 \text{ mmol HNO}_3 \times \frac{1.00 \text{ mL}}{0.101 \text{ mmol HNO}_3} = 50.1 \text{ mL HNO}_3 \text{ required}$$

d. $KOH(aq) + HNO_3(aq) \rightarrow KNO_3(aq) + H_2O(l)$

$$1.21 \text{ L} \times \frac{0.102 \text{ mol}}{1.00 \text{ L}} = 0.123 \text{ mol KOH present in the sample}$$

$$0.123 \text{ mol KOH} \times \frac{1 \text{ mol HNO}_3}{1 \text{ mol KOH}} = 0.123 \text{ mol HNO}_3 \text{ required}$$

$$0.123 \text{ mol HNO}_3 \times \frac{1.00 \text{ L}}{0.101 \text{ mol HNO}_3} = 1.22 \text{ L HNO}_3 \text{ required}$$

76. 1 normal

78. 1.53 equivalents OH^- ion are needed to react with 1.53 equivalents of H^+ ion. By *definition*, one equivalent of OH^- ion exactly neutralizes one equivalent of H^+ ion.

80. $N = \dfrac{\text{number of equivalents of solute}}{\text{number of liters of solution}}$

a. equivalent weight NaOH = molar mass NaOH = 40.00 g

$$0.113 \text{ g NaOH} \times \frac{1 \text{ equiv NaOH}}{40.00 \text{ g}} = 2.83 \times 10^{-3} \text{ equiv NaOH}$$

$$10.2 \text{ mL} = 0.0102 \text{ L}$$

$$N = \frac{2.83 \times 10^{-3} \text{ equiv}}{0.0102 \text{ L}} = 0.277 \, N$$

b. equivalent weight $Ca(OH)_2 = \dfrac{\text{molar mass}}{2} = \dfrac{74.10 \text{ g}}{2} = 37.05 \text{ g}$

$$12.5 \text{ mg} \times \frac{1 \text{ g}}{10^3 \text{ mg}} \times \frac{1 \text{ equiv}}{37.05 \text{ g}} = 3.37 \times 10^{-4} \text{ equiv } Ca(OH)_2$$

$$100. \text{ mL} = 0.100 \text{ L}$$

$$N = \frac{3.37 \times 10^{-3} \text{ equiv}}{0.100 \text{ L}} = 3.37 \times 10^{-3} \, N$$

c. equivalent weight $H_2SO_4 = \dfrac{\text{molar mass}}{2} = \dfrac{98.09 \text{ g}}{2} = 49.05 \text{ g}$

$$12.4 \text{ g} \times \frac{1 \text{ equiv}}{49.05 \text{ g}} = 0.253 \text{ equiv } H_2SO_4$$

$$155 \text{ mL} = 0.155 \text{ L}$$

$$N = \frac{0.253 \text{ equiv}}{0.155 \text{ L}} = 1.63 \, N$$

82. a. $0.134 \, M$ NaOH $\times \dfrac{1 \text{ equiv NaOH}}{1 \text{ mol NaOH}} = 0.134 \, N$ NaOH

    b. $0.00521 \, M$ $Ca(OH)_2 \times \dfrac{2 \text{ equiv } Ca(OH)_2}{1 \text{ mol } Ca(OH)_2} = 0.0104 \, N$ $Ca(OH)_2$

    c. $4.42 \, M$ $H_3PO_4 \times \dfrac{3 \text{ equiv } H_3PO_4}{1 \text{ mol } H_3PO_4} = 13.3 \, N$ H_3PO_4

84. Molar mass of $Ca(OH)_2 = 74.10$ g

    $$5.21 \text{ mg } Ca(OH)_2 \times \frac{1 \text{ g}}{10^3 \text{ mg}} \times \frac{1 \text{ mol}}{74.10 \text{ g}} = 7.03 \times 10^{-5} \text{ mol } Ca(OH)_2$$

    1000. mL = 1.000 L (volumetric flask volume: 4 significant figures).

    $$M = \frac{7.03 \times 10^{-5} \text{ mol}}{1.000 \text{ L}} = 7.03 \times 10^{-5} \, M \, Ca(OH)_2$$

    $$N = 7.03 \times 10^{-5} \, M \, Ca(OH)_2 \times \frac{2 \text{ equiv } Ca(OH)_2}{1 \text{ mol } Ca(OH)_2} = 1.41 \times 10^{-4} \, N \, Ca(OH)_2$$

86. $H_2SO_4(aq) + 2NaOH \rightarrow Na_2SO_4(aq) + 2H_2O(l)$

    $0.145 \, M$ NaOH = $0.145 \, N$ NaOH $56.2 \text{ mL} = 0.0562 \text{ L}$

186 Chapter 14   Solutions

$$0.0562 \text{ L NaOH} \times \frac{0.145 \text{ equiv}}{1.00 \text{ L}} = 0.00815 \text{ equiv NaOH}$$

0.00815 equiv NaOH requires 0.00815 equiv H_2SO_4 to react.

$$0.00815 \text{ equiv } H_2SO_4 \times \frac{1.00 \text{ L}}{0.172 \text{ equiv}} = 0.0474 \text{ L} = 47.4 \text{ mL } H_2SO_4 \text{ solution}$$

88. $2\text{NaOH}(aq) + H_2SO_4(aq) \rightarrow Na_2SO_4(aq) + 2 H_2O(l)$

$$27.34 \text{ mL NaOH} \times \frac{0.1021 \text{ mmol}}{1.00 \text{ mL}} = 2.791 \text{ mmol NaOH}$$

$$2.791 \text{ mmol NaOH} \times \frac{1 \text{ mmol } H_2SO_4}{2 \text{ mmol NaOH}} = 1.396 \text{ mmol } H_2SO_4$$

$$M = \frac{1.396 \text{ mmol } H_2SO_4}{25.00 \text{ mL}} = 0.05583 \text{ } M \text{ } H_2SO_4 = 0.1117 \text{ } N \text{ } H_2SO_4$$

90. Molarity is defined as the number of moles of solute contained in 1 liter of *total* solution volume (solute plus solvent after mixing). In the first case, where 50. g of NaCl is dissolved in 1.0 L of water, the total volume after mixing is *not* known and the molarity cannot be calculated. In the second example, the final volume after mixing is known and the molarity can be calculated simply.

92. $$75 \text{ g solution} \times \frac{25 \text{ g NaCl}}{100 \text{ g solution}} = 18.75 \text{ g NaCl}$$

$$\text{new \%} = \frac{18.75 \text{ g NaCl}}{575 \text{ g solution}} \times 100 = 3.26 = 3.3 \text{ \%}$$

94. $Ba(NO_3)_2(aq) + H_2SO_4(aq) \rightarrow BaSO_4(s) + 2HNO_3(aq)$

37.5 mL = 0.0375 L

$$0.0375 \text{ L} \times \frac{0.221 \text{ mol } H_2SO_4}{1.00 \text{ L}} = 0.00829 \text{ mol } H_2SO_4$$

Since the coefficients of $Ba(NO_3)_2$ and H_2SO_4 in the balanced chemical equation for the reaction are both *one*, then 0.00829 mol of Ba^{2+} ion will be precipitated from the solution as $BaSO_4$.

molar mass $BaSO_4$ = 233.4 g

$$0.00829 \text{ mol } BaSO_4 \times \frac{233.4 \text{ g } BaSO_4}{1 \text{ mol } BaSO_4} = 1.93 \text{ g } BaSO_4 \text{ precipitate}$$

96. molar mass H_2O = 18.0 g

1.0 L water = 1.0×10^3 mL water ≈ 1.0×10^3 g water

$$1.0 \times 10^3 \text{ g } H_2O \times \frac{1 \text{ mol } H_2O}{18.0 \text{ g } H_2O} = 56 \text{ mol } H_2O$$

Chapter 14 Solutions

98. 500 mL HCl solution = 0.500 L HCl solution

$$0.500 \text{ L solution} \times \frac{0.100 \text{ mol HCl}}{1.00 \text{ L HCl solution}} = 0.0500 \text{ mol HCl}$$

$$0.0500 \text{ mol HCl} \times \frac{22.4 \text{ L HCl gas at STP}}{1 \text{ mol HCl}} = 1.12 \text{ L HCl gas at STP}$$

100. $10.0 \text{ g HCl} \times \dfrac{100. \text{ g solution}}{33.1 \text{ g HCl}} = 30.21 \text{ g solution}$

$$30.21 \text{ g solution} \times \frac{1.00 \text{ mL solution}}{1.147 \text{ g solution}} = 26.3 \text{ mL solution}$$

102. molar mass $CaCl_2$ = 111.0 g

$$14.2 \text{ g } CaCl_2 \times \frac{1 \text{ mol } CaCl_2}{111.0 \text{ g } CaCl_2} = 0.128 \text{ mol } CaCl_2$$

50.0 mL = 0.0500 L

$$M = \frac{0.128 \text{ mol } CaCl_2}{0.0500 \text{ L}} = 2.56 \text{ } M$$

104. a. $\dfrac{5.0 \text{ g } KNO_3}{(5.0 \text{ g } KNO_3 + 75 \text{ g } H_2O)} \times 100 = \dfrac{5.0 \text{ g}}{80.0 \text{ g}} \times 100 = 6.3\% \text{ } KNO_3$

b. 2.5 mg = 0.0025 g

$$\frac{0.0025 \text{ g } KNO_3}{(0.0025 \text{ g } KNO_3 + 1.0 \text{ g } H_2O)} \times 100 = \frac{0.0025 \text{ g}}{1.0025 \text{ g}} \times 100 = 0.25\% \text{ } KNO_3$$

c. $\dfrac{11 \text{ g } KNO_3}{(11 \text{ g } KNO_3 + 89 \text{ g } H_2O)} \times 100 = \dfrac{11 \text{ g}}{100 \text{ g}} \times 100 = 11\% \text{ } KNO_3$

d. $\dfrac{11 \text{ g } KNO_3}{(11 \text{ g } KNO_3 + 49 \text{ g } H_2O)} \times 100 = \dfrac{11 \text{ g}}{60 \text{ g}} \times 100 = 18\% \text{ } KNO_3$

106. $\%C = \dfrac{5.0 \text{ g C}}{(5.0 \text{ g C} + 1.5 \text{ g Ni} + 100. \text{ g Fe})} \times 100 = \dfrac{5.0 \text{ g}}{106.5 \text{ g}} \times 100 = 4.7\% \text{ C}$

$\%Ni = \dfrac{1.5 \text{ g Ni}}{(5.0 \text{ g C} + 1.5 \text{ g Ni} + 100. \text{ g Fe})} \times 100 = \dfrac{1.5 \text{ g}}{106.5 \text{ g}} \times 100 = 1.4\% \text{ Ni}$

$\%Fe = \dfrac{100. \text{ g Fe}}{(5.0 \text{ g C} + 1.5 \text{ g Ni} + 100. \text{ g Fe})} \times 100 = \dfrac{100. \text{ g}}{106.5 \text{ g}} \times 100 = 93.9\% \text{ Fe}$

188 Chapter 14 Solutions

108. To say that the solution is 5.5% by mass Na_2CO_3 means that 5.5 g of Na_2CO_3 are contained in every 100 g of the solution.

$$500.\text{ g solution} \times \frac{5.5\text{ g }Na_2CO_3}{100\text{ g solution}} = 28\text{ g }Na_2CO_3$$

110. For NaCl: $125\text{ g solution} \times \dfrac{7.5\text{ g NaCl}}{100\text{ g solution}} = 9.4\text{ g NaCl}$

For KBr: $125\text{ g solution} \times \dfrac{2.5\text{ g KBr}}{100\text{ g solution}} = 3.1\text{ g KBr}$

112. Molarity = $\dfrac{\text{moles of solute}}{\text{liters of solution}}$

a. 25 mL = 0.025 L

$$M = \frac{0.10\text{ mol }CaCl_2}{0.025\text{ L solution}} = 4.0\ M$$

b. $M = \dfrac{2.5\text{ mol KBr}}{2.5\text{ L solution}} = 1.0\ M$

c. 755 mL = 0.755 L

$$M = \frac{0.55\text{ mol }NaNO_3}{0.755\text{ L solution}} = 0.73\ M$$

d. $M = \dfrac{4.5\text{ mol }Na_2SO_4}{1.25\text{ L solution}} = 3.6\ M$

114. molar mass $C_{12}H_{22}O_{11}$ = 342.3 g

$$125\text{ g }C_{12}H_{22}O_{11} \times \frac{1\text{ mol}}{342.3\text{ g}} = 0.3652\text{ mol }C_{12}H_{22}O_{11}$$

450. mL = 0.450 L

$$M = \frac{0.3652\text{ mol }C_{12}H_{22}O_{11}}{0.450\text{ L solution}} = 0.812\ M$$

116. molar mass NaCl = 58.44 g

$$1.5\text{ g NaCl} \times \frac{1\text{ mol}}{58.44\text{ g}} = 0.0257\text{ mol NaCl}$$

$$M = \frac{0.0257\text{ mol NaCl}}{1.0\text{ L solution}} = 0.026\ M$$

118. a. $4.25\text{ L solution} \times \dfrac{0.105\text{ mol KCl}}{1.00\text{ L solution}} = 0.446\text{ mol KCl}$

molar mass KCl = 74.6 g

$$0.446 \text{ mol KCl} \times \frac{74.6 \text{ g KCl}}{1 \text{ mol KCl}} = 33.3 \text{ g KCl}$$

b. 15.1 mL = 0.0151 L

$$0.0151 \text{ L solution} \times \frac{0.225 \text{ mol NaNO}_3}{1.00 \text{ L solution}} = 3.40 \times 10^{-3} \text{ mol NaNO}_3$$

molar mass $NaNO_3$ = 85.00 g

$$3.40 \times 10^{-3} \text{ mol} \times \frac{85.00 \text{ g NaNO}_3}{1 \text{ mol NaNaNO}_3} = 0.289 \text{ g NaNO}_3$$

c. 25 mL = 0.025 L

$$0.025 \text{ L solution} \times \frac{3.0 \text{ mol HCl}}{1.00 \text{ L solution}} = 0.075 \text{ mol HCl}$$

molar mass HCl = 36.46 g

$$0.075 \text{ mol HCl} \times \frac{36.46 \text{ g HCl}}{1 \text{ mol HCl}} = 2.7 \text{ g HCl}$$

d. 100. mL = 0.100 L

$$0.100 \text{ L solution} \times \frac{0.505 \text{ mol H}_2\text{SO}_4}{1.00 \text{ L solution}} = 0.0505 \text{ mol H}_2\text{SO}_4$$

molar mass H_2SO_4 = 98.09 g

$$0.0505 \text{ mol H}_2\text{SO}_4 \times \frac{98.09 \text{ g H}_2\text{SO}_4}{1 \text{ mol H}_2\text{SO}_4} = 4.95 \text{ g H}_2\text{SO}_4$$

120. a.
$$1.25 \text{ L} \times \frac{0.250 \text{ mol Na}_3\text{PO}_4}{1.00 \text{ L}} = 0.3125 \text{ mol Na}_3\text{PO}_4$$

$$0.3125 \text{ mol Na}_3\text{PO}_4 \times \frac{3 \text{ mol Na}^+}{1 \text{ mol Na}_3\text{PO}_4} = 0.938 \text{ mol Na}^+$$

$$0.3125 \text{ mol Na}_3\text{PO}_4 \times \frac{1 \text{ mol PO}_4^{3-}}{1 \text{ mol Na}_3\text{PO}_4} = 0.313 \text{ mol PO}_4^{3-}$$

b. 3.5 mL = 0.0035 L

$$0.0035 \text{ L} \times \frac{6.0 \text{ mol H}_2\text{SO}_4}{1.00 \text{ L}} = 0.021 \text{ mol H}_2\text{SO}_4$$

$$0.021 \text{ mol H}_2\text{SO}_4 \times \frac{2 \text{ mol H}^+}{1 \text{ mol H}_2\text{SO}_4} = 0.042 \text{ mol H}^+$$

$$0.021 \text{ mol H}_2\text{SO}_4 \times \frac{1 \text{ mol SO}_4^{2-}}{1 \text{ mol H}_2\text{SO}_4} = 0.021 \text{ mol SO}_4^{2-}$$

190   Chapter 14   Solutions

c.   25 mL = 0.025 L

$$0.025 \text{ L} \times \frac{0.15 \text{ mol AlCl}_3}{1.00 \text{ L}} = 0.00375 \text{ mol AlCl}_3$$

$$0.00375 \text{ mol AlCl}_3 \times \frac{1 \text{ mol Al}^{3+}}{1 \text{ mol AlCl}_3} = 0.0038 \text{ mol Al}^{3+}$$

$$0.00375 \text{ mol AlCl}_3 \times \frac{3 \text{ mol Cl}^-}{1 \text{ mol AlCl}_3} = 0.011 \text{ mol Cl}^-$$

d. $1.50 \text{ L} \times \dfrac{1.25 \text{ mol BaCl}_2}{1.00 \text{ L}} = 1.875 \text{ mol BaCl}_2$

$$1.875 \text{ mol BaCl}_2 \times \frac{1 \text{ mol Ba}^{2+}}{1 \text{ mol BaCl}_2} = 1.88 \text{ mol Ba}^{2+}$$

$$1.875 \text{ mol BaCl}_2 \times \frac{2 \text{ mol Cl}^-}{1 \text{ mol BaCl}_2} = 3.75 \text{ mol Cl}^-$$

122. $M_1 \times V_1 = M_2 \times V_2$

a. $M_1 = 0.200 \ M$ $M_2 = ?$
     $V_1 = 125 \text{ mL}$ $V_2 = 125 + 150. = 275 \text{ mL}$

$$M_2 = \frac{(0.200 \ M)(125 \text{ mL})}{(275 \text{ mL})} = 0.0909 \ M$$

b. $M_1 = 0.250 \ M$ $M_2 = ?$
     $V_1 = 155 \text{ mL}$ $V_2 = 155 + 150. = 305 \text{ mL}$

$$M_2 = \frac{(0.250 \ M)(155 \text{ mL})}{(305 \text{ mL})} = 0.127 \ M$$

c. $M_1 = 0.250 \ M$ $M_2 = ?$
     $V_1 = 0.500 \text{ L} = 500. \text{ mL}$ $V_2 = 500. + 150. = 650. \text{ mL}$

$$M_2 = \frac{(0.250 \text{ M})(500. \text{ mL})}{(650 \text{ mL})} = 0.192 \ M$$

d. $M_1 = 18.0 \ M$ $M_2 = ?$
     $V_1 = 15 \text{ mL}$ $V_2 = 15 + 150. = 165 \text{ mL}$

$$M_2 = \frac{(18.0 \ M)(15 \text{ mL})}{(165 \text{ mL})} = 1.6 \ M$$

124. $M_1 \times V_1 = M_2 \times V_2$
      $M_1 = 5.4 \ M$ $M_2 = ?$
      $V_1 = 50. \text{ mL}$ $V_2 = 300. \text{ mL}$

$$M_2 = \frac{(5.4 \ M)(50. \ \text{mL})}{(300. \ \text{mL})} = 0.90 \ M$$

126. 25.0 mL = 0.0250 L

$$0.0250 \ \text{L NiCl}_2 \ \text{solution} \times \frac{0.20 \ \text{mol NiCl}_2}{1.00 \ \text{L NiCl}_2 \ \text{solution}} = 0.00500 \ \text{mol NiCl}_2$$

$$0.00500 \ \text{mol NiCl}_2 \times \frac{1 \ \text{mol Na}_2\text{S}}{1 \ \text{mol NiCl}_2} = 0.00500 \ \text{mol Na}_2\text{S}$$

$$0.00500 \ \text{mol Na}_2\text{S} \times \frac{1.00 \ \text{L Na}_2\text{S solution}}{0.10 \ \text{mol Na}_2\text{S}} = 0.050 \ \text{L} = 50. \ \text{mL Na}_2\text{S solution}$$

128. $HNO_3(aq) + NaOH(aq) \rightarrow NaNO_3(aq) + H_2O(l)$

35.0 mL = 0.0350 L

$$0.0350 \ \text{L} \times \frac{0.150 \ \text{mol NaOH}}{1.00 \ \text{L}} = 5.25 \times 10^{-3} \ \text{mol NaOH}$$

$$5.25 \times 10^{-3} \ \text{mol NaOH} \times \frac{1 \ \text{mol HNO}_3}{1 \ \text{mol NaOH}} = 5.25 \times 10^{-3} \ \text{mol HNO}_3$$

$$5.25 \times 10^{-3} \ \text{mol HNO}_3 \times \frac{1.00 \ \text{L}}{0.150 \ \text{mol HNO}_3} = 0.0350 \ \text{L} = 35.0 \ \text{mL HNO}_3$$

130. $N = \dfrac{\text{number of equivalents of solute}}{\text{number of liters of solution}}$

    a.   equivalent weight HCl = molar mass HCl = 36.46 g

$$15.0 \ \text{g HCl} \times \frac{1 \ \text{equiv HCl}}{36.46 \ \text{g HCl}} = 0.411 \ \text{equiv HCl}$$

500. mL = 0.500 L

$$N = \frac{0.411 \ \text{equiv}}{0.500 \ \text{L}} = 0.822 \ N$$

    b.   equivalent weight $H_2SO_4 = \dfrac{\text{molar mass}}{2} = \dfrac{98.09 \ \text{g}}{2} = 49.05 \ \text{g}$

$$49.0 \ \text{g H}_2\text{SO}_4 \times \frac{1 \ \text{equiv H}_2\text{SO}_4}{49.05 \ \text{g H}_2\text{SO}_4} = 0.999 \ \text{equiv H}_2\text{SO}_4$$

250. mL = 0.250 L

$$N = \frac{0.999 \ \text{equiv}}{0.250 \ \text{L}} = 4.00 \ N$$

    c.   equivalent weight $H_3PO_4 = \dfrac{\text{molar mass}}{3} = \dfrac{98.0 \ \text{g}}{3} = 32.67 \ \text{g}$

192 Chapter 14 Solutions

$$10.0 \text{ g H}_3\text{PO}_4 \times \frac{1 \text{ equiv H}_3\text{PO}_4}{32.67 \text{ g H}_3\text{PO}_4} = 0.3061 \text{ equiv H}_3\text{PO}_4$$

$$100. \text{ mL} = 0.100 \text{ L}$$

$$N = \frac{0.3061 \text{ equiv}}{0.100 \text{ L}} = 3.06 \text{ } N$$

132. molar mass NaH_2PO_4 = 120.0 g

$$5.0 \text{ g NaH}_2\text{PO}_4 \times \frac{1 \text{ mol NaH}_2\text{PO}_4}{120.0 \text{ g NaH}_2\text{PO}_4} = 0.04167 \text{ mol NaH}_2\text{PO}_4$$

$$500. \text{ mL} = 0.500 \text{ L}$$

$$M = \frac{0.04167 \text{ mol}}{0.500 \text{ L}} = 0.08333 \text{ } M \text{ NaH}_2\text{PO}_4 = 0.083 \text{ } M \text{ NaH}_2\text{PO}_4$$

$$0.08333 \text{ } M \text{ NaH}_2\text{PO}_4 \times \frac{2 \text{ equiv NaH}_2\text{PO}_4}{1 \text{ mol NaH}_2\text{PO}_4} = 0.1667 \text{ } N \text{ NaH}_2\text{PO}_4 = 0.17 \text{ } N \text{ NaH}_2\text{PO}_4$$

134. $N_{acid} \times V_{acid} = N_{base} \times V_{base}$

$N_{acid} \times (10.0 \text{ mL}) = (3.5 \times 10^{-2} \text{ } N)(27.5 \text{ mL})$

$N_{acid} = 9.6 \times 10^{-2} \text{ } N \text{ HNO}_3$

Cumulative Review: Chapters 12, 13, and 14

2. The pressure exerted by the atmosphere is due to the several mile thick layer of gases above the surface of the earth pressing down on us. Atmospheric pressure has traditionally been measured with a mercury barometer (see Figure 12.2). A mercury barometer usually consists of a glass tube which is sealed at one end and filled with mercury. The tube is then inverted over an open reservoir also containing mercury. When the tube is inverted, most of the mercury does not fall out of the tube. Since the reservoir of mercury is open to the atmosphere, the atmospheric pressure on the surface of the mercury in the reservoir is enough pressure to hold the bulk of the mercury in the glass tube. The pressure of the atmosphere is sufficient, on average, to support a column of mercury 76 cm (760 mm) high in the tube.

4. Boyle's law basically says that the volume of a gas sample will decrease if you squeeze harder on it. Imagine squeezing hard on a tennis ball with your hand: the ball collapses as the gas inside it is forced into a smaller volume by your hand. Of course, to be perfectly correct, the temperature and amount of gas (moles) must remain the same while you adjust the pressure for Boyle's law to hold true. The first of the two mathematical statements of Boyle's law you should remember is

   $P \times V = \text{constant}$

   which basically is the definition of Boyle's law (in order for the product ($P \times V$) to remain constant, if one of these terms increases the other must decrease). The second formulation of Boyle's law you have to be able to deal with is the one more commonly used in solving problems,

   $P_1 \times V_1 = P_2 \times V_2$

   With this second formulation, we can determine pressure-volume information about a given sample under two sets of conditions. These two mathematical formulas are just two different ways of saying the same thing: if the pressure on a sample of gas is increased, the volume of the sample of gas will decrease. A graph of Boyle's law data is given as Figure 12.5: this sort of graph ($xy = k$) is known to mathematicians as a hyperbola.

6. Charles's law basically says that if you heat a sample of gas, the volume of the sample will increase. That is, when the temperature of a gas is increased, the volume of the gas also increases (assuming the pressure and amount of gas remains the same). Charles's law is a direct proportionality when the temperature is expressed in kelvins (if you increase T, this increases V), whereas Boyle's law is an inverse proportionality (if you increase P, this decreases V). There are two mathematical statements of Charles's law you should be familiar with. The first statement

   $V = bT$

   is just a definition (the volume of a gas sample is directly related to its Kelvin temperature: if you increase the temperature, the volume increases). The working formulation of Charles's law we use in problem solving is given as

$$\frac{V_1}{T_1} = \frac{V_2}{T_2}$$

With this formulation, we can determine volume-temperature information for a given gas sample under two sets of conditions. Charles's law only holds true if the amount of gas remains the same (obviously the volume of a gas sample would increase if there were more gas present) and also if the pressure remains the same (a change in pressure also changes the volume of a gas sample).

8. Avogadro's law tells us that, with all other things being equal, two moles of gas is twice as big as one mole of gas! That is, the volume of a sample of gas is directly proportional to the number of moles or molecules of gas present (at constant temperature and pressure). If we want to compare the volumes of two samples of the same gas as an indication of the amount of gas present in the samples, we would have to make certain that the two samples of gas are at the same pressure and temperature: the volume of a sample of gas would vary with either temperature or pressure, or both. Avogadro's law holds true for comparing gas samples that are under the same conditions. Avogadro's law is a direct proportionality: the greater the number of gas molecules you have in a sample, the larger the sample's volume will be.

10. The "partial" pressure of an individual gas in a mixture of gases represents the pressure the gas would have in the same container at the same temperature if it were the only gas present. The total pressure in a mixture of gases is just the sum of the individual partial pressures of the gases present in the mixture. Because the partial pressures of the gases in a mixture are additive (i.e., the total pressure is the sum of the partial pressures), this suggests that the total pressure in a container is a function really only of the number of molecules present in the same, and not of the identity of the molecules or any other property of the molecules (such as their inherent atomic size).

12. The main postulates of the kinetic-molecular theory for gases are as follows: (a) gases consist of tiny particles (atoms or molecules), and the size of these particles themselves is negligible compared to the bulk volume of a gas sample; (b) the particles in a gas are in constant random motion, colliding with each other and with the walls of the container; (c) the particles in a gas sample do not exert any attractive or repulsive forces on one another; (d) the average kinetic energy of the particles in a sample of gas is directly related to the absolute temperature of the gas sample. The pressure exerted by a gas is a result of the molecules colliding with (and pushing on) the walls of the container; the pressure increases with temperature because at a higher temperature, the molecules are moving faster and hit the walls of the container with greater force. A gas fills whatever volume is available to it because the molecules in a gas are in constant random motion: if

the motion of the molecules is random, they eventually will move out into whatever volume is available until the distribution of molecules is uniform; at constant pressure, the volume of a gas sample increases as the temperature is increased because with each collision having greater force, the container must expand so that the molecules (and therefore the collisions) are farther apart if the pressure is to remain constant.

14. Solids and liquids are much more condensed states of matter than are gases: the molecules are much closer together in solids and liquids and interact with each other to a much greater extent. Solids and liquids have much greater densities than do gases, and are much less compressible, because there is so little room between the molecules in the solid and liquid states (solids and liquids effectively have native volumes of their own, and their volumes are not affected nearly as much by the temperature or pressure). Although solids are more rigid than liquids, the solid and liquid state have much more in common with each other than either of these states has with the gaseous state. We know this is true since it typically only takes a few kilojoules of energy to melt 1 mol of a solid (since not much change has to take place in the molecules), whereas it may take 10 times more energy to vaporize a liquid (since there is such a great change between the liquid and gaseous states).

16. The normal boiling point of water, that is, water's boiling point at a pressure of exactly 760 mm Hg, is 100°C (you will recall that the boiling point of water was used to set one of the reference temperatures of the Celsius temperature scale). Water remains at 100°C while boiling, until all the water has boiled away, because the additional heat energy being added to the sample is used to overcome attractive forces among the water molecules as they go from the condensed, liquid state to the gaseous state. The normal (760 mm Hg) freezing point of water is exactly 0°C (again, this property of water was used as one of the reference points for the Celsius temperature scale). A cooling curve for water is given as Figure 13.2: notice how the curve shows that the amount of heat needed to boil the sample is much larger than the amount needed to melt the sample.

18. Dipole-dipole forces are a type of intermolecular force that can exist between molecules with permanent dipole moments. Molecules with permanent dipole moments try to orient themselves so that the positive end of one polar molecule can attract the negative end of another polar molecule. Dipole-dipole forces are not nearly as strong as ionic or covalent bonding forces (only about 1% strong as covalent bonding forces) since electrostatic attraction is related to the magnitude of the charges of the attracting species. Since polar molecules have only a "partial" charge at each end of the dipole, the magnitude of the attractive force is not as large. The strength of such forces also drops rapidly as molecules become farther apart and is important only in the solid and liquid states (such forces are negligible in the gaseous state since the molecules are too far apart). Hydrogen bonding is an especially strong sort of dipole-dipole attractive force which can exist when hydrogen atoms are directly bonded to the most strongly

electronegative atoms (N, O, and F). Because the hydrogen atom is so small, dipoles involving N-H, O-H, and F-H bonds can approach each other much more closely than can dipoles involving other atoms. Since the magnitude of dipole-dipole forces is dependent on distance, unusually strong attractive forces can exist in such molecules. We take the fact that the boiling point of water is so much higher than that of the other covalent hydrogen compounds of the Group 6 elements as evidence for the special strength of hydrogen bonding (it takes more energy to vaporize water because of the extra strong forces holding together the molecules in the liquid state).

20. Vaporization of a liquid requires and input of energy because the intermolecular forces which hold the molecules together in the liquid state must be overcome. The high heat of vaporization of water is essential to life on earth since much of the excess energy striking the earth from the sun is dissipated in vaporizing water. Condensation is the opposite process to vaporization: that is, condensation refers to the process by which molecules in the vapor state form a liquid. In a closed container containing a liquid and some empty space above the liquid, an equilibrium is set up between vaporization and condensation. The liquid in such a sealed container never completely evaporates: when the liquid is first placed in the container, the liquid phase begins to evaporate into the empty space. As the number of molecules in the vapor phase begins to get large, however, some of these molecules begin to re-enter the liquid phase. Eventually, every time a molecule of liquid somewhere in the container enters the vapor phase, somewhere else in the container a molecule of vapor re-enters the liquid. There is no further net change in the amount of liquid phase (although molecules are continually moving between the liquid and vapor phases). The pressure of the vapor in such an equilibrium situation is characteristic for the liquid at each particular temperature (for example, the vapor pressures of water are tabulated at different temperatures in Table 12.2). A simple experiment to determine vapor pressure is shown in Figure 13.10: samples of a liquid are injected into a sealed tube containing mercury; since mercury is so dense, the liquids float to the top of the mercury where they evaporate; as the vapor pressures of the liquids develop to the saturation point, the level of mercury in the tube changes as an index of the magnitude of the vapor pressures. Typically, liquids with strong intermolecular forces have small vapor pressures (they have more difficulty in evaporating) than do liquids with very weak intermolecular forces: for example, the components of gasoline (weak forces) have much higher vapor pressures than does water (strong forces) and evaporate more easily.

22. The simple model we use to explain many properties of metallic elements is called the electron sea model. In this model we picture a regular lattice array of metal cations in sort of a "sea" of mobile valence electrons. The electrons can move easily to conduct heat or electricity through the metal, and the lattice of cations can be deformed fairly easily, allowing the metal to be hammered into a sheet or stretched to make a wire. An alloy is a material that contains a mixture of elements, which overall has metallic properties. Substitutional alloys consist of

a host metal in which some of the atoms in the metal's crystalline structure are replaced by atoms of other metallic elements of comparable size to the atoms of the host metal. For example, sterling silver consists of an alloy in which approximately 7% of the silver atoms have been replaced by copper atoms. Brass and pewter are also substitutional alloys. An interstitial alloy is formed when other smaller atoms enter the interstices (holes) between atoms in the host metal's crystal structure. Steel is an interstitial alloy in which typically carbon atoms enter the interstices of a crystal of iron atoms. The presence of the interstitial carbon atoms markedly changes the properties of the iron, making it much harder, more malleable, and more ductile. Depending on the amount of carbon introduced into the iron crystals, the properties of the steel resulting can be carefully controlled.

24. A saturated solution is one that contains as much solute as can dissolve at a particular temperature. To say that a solution is saturated doe not necessarily mean that the solute is present at a high concentration: for example, magnesium hydroxide only dissolves to a very small extent before the solution is saturated, whereas it takes a great deal of sugar to form a saturated solution (and the saturated solution is extremely concentrated). A saturated solution is one which is in equilibrium with undissolved solute: as molecules of solute dissolve from the solid in one place in the solution, dissolved molecules rejoin the solid phase in another place in the solution. As with the development of vapor pressure above a liquid (see Question 20 above), formation of a solution reaches a state of dynamic equilibrium: once the rates of dissolving and "undissolving" become equal, there will be no further net change in the concentration of the solution and the solution will be saturated.

26. Adding additional solvent to a solution so as to dilute the solution *does not change* the number of moles of solute present, but only changes the volume in which the solute is dispersed. If we are using the molarity of the solution to describe its concentration, the number of liters is changed when we add solvent, and the number of moles per liter (the molarity) changes, but the actual number of moles of solute does not change. For example, 125 mL of 0.551 M NaCl contains 68.9 millimol of NaCl. The solution will still contain 68.9 millimol of NaCl after the 250 mL of water is added to it, only now the 68.9 millimol of NaCl will be dispersed in a total volume of 375 mL. This gives the new molarity as 68.9 mmol/375 mL = 0.184 M. The volume and the concentration have changed, but the number of moles of solute in the solution has not changed.

28. $$V_2 = \frac{P_1 \times V_1}{P_2}$$

   a. 2.41 atm = 1.83×10^3 mm Hg

   $$V_2 = \frac{(759 \text{ mm Hg})(245 \text{ mL})}{(1.83 \times 10^3 \text{ mm Hg})} = 102 \text{ mL}$$

Cumulative Review: Chapters 12, 13, and 14

    b. $\quad V_2 = \dfrac{(759 \text{ mm Hg})(2.71 \text{ L})}{(1104 \text{ mmHg})} = 1.86 \text{ L}$

    c. $\quad 204 \text{ kPa} = 1530 \text{ mm Hg}$

$$V_2 = \dfrac{(759 \text{ mm Hg})(45.2 \text{ mL})}{(1530 \text{ mm Hg})} = 22.4 \text{ mL}$$

30. a. $P_1 = 775$ mm Hg $\qquad P_2 = 760$ mm Hg

        $V_1 = 45.1$ mL $\qquad\qquad V_2 = ?$

        $T_1 = 24.1°C = 297$ K $\qquad T_2 = 273$ K

$$V_2 = \dfrac{P_1 V_1 T_2}{T_1 P_2} = \dfrac{(775 \text{ mm Hg})(45.1 \text{ mL})(273 \text{ K})}{(297 \text{ K})(760 \text{ mm Hg})} = 42.3 \text{ mL}$$

    b. $P_1 = 0.890$ atm $\qquad P_2 = 1.00$ atm

        $V_1 = 4.31$ L $\qquad\qquad V_2 = ?$

        $T_1 = 72.1°C = 345$ K $\qquad T_2 = 273$ K

$$V_2 = \dfrac{P_1 V_1 T_2}{T_1 P_2} = \dfrac{(0.890 \text{ atm})(4.31 \text{ L})(273 \text{ K})}{(345 \text{ K})(1.00 \text{ atm})} = 3.03 \text{ L}$$

    c. $P_1 = 91.2$ kPa $\qquad P_2 = 101.325$ kPa

        $V_1 = 5.12$ mL $\qquad\qquad V_2 = ?$

        $T_1 = 289$ K $\qquad\qquad T_2 = 273$ K

$$V_2 = \dfrac{P_1 V_1 T_2}{T_1 P_2} = \dfrac{(91.2 \text{ kPa})(5.12 \text{ mL})(273 \text{ K})}{(289 \text{ K})(101.325 \text{ kPa})} = 4.35 \text{ mL}$$

    d. $P_1 = 1.45$ atm $\qquad P_2 = 1.00$ atm

        $V_1 = 91.3$ L $\qquad\qquad V_2 = ?$

        $T_1 = 451°C = 724$ K $\qquad T_2 = 273$ K

$$V_2 = \dfrac{P_1 V_1 T_2}{T_1 P_2} = \dfrac{(1.45 \text{ atm})(91.3 \text{ L})(273 \text{ K})}{(724 \text{ K})(1.00 \text{ atm})} = 49.9 \text{ L}$$

32. Since we have 100. g of the solution, which is 3.11% by mass H_2O_2, then the sample must contain 3.11 g of H_2O_2

    molar mass H_2O_2 = 34.02 g

Cumulative Review: Chapters 12, 13, and 14   199

$$3.11 \text{ g H}_2\text{O}_2 \times \frac{1 \text{ mol H}_2\text{O}_2}{34.02 \text{ g H}_2\text{O}_2} = 0.0914 \text{ mol H}_2\text{O}_2$$

$$0.0914 \text{ mol H}_2\text{O}_2 \times \frac{1 \text{ mol O}_2}{2 \text{ mol H}_2\text{O}_2} = 0.0457 \text{ mol O}_2 \text{ produced}$$

At STP, $0.0457 \text{ mol O}_2 \times \dfrac{22.4 \text{ L O}_2}{1 \text{ mol O}_2} = 1.02 \text{ L O}_2$

$P_1 = 760$ mm Hg $P_2 = 771$ mm Hg

$V_1 = 1.02$ L $V_2 = ?$

$T_1 = 273$ K $T_2 = 297$ K

$$V_2 = \frac{P_1 V_1 T_2}{T_1 P_2} = \frac{(760 \text{ mm Hg})(1.02 \text{ L})(297 \text{ K})}{(273 \text{ K})(771 \text{ mm Hg})} = 1.09 \text{ L}$$

34. a. mass of solution = 4.25 + 7.50 + 52.0 = 63.75 g (63.8 g)

$\dfrac{4.25 \text{ g NaCl}}{63.75 \text{ g}} \times 100 = 6.67 \text{ \%NaCl}$ $\dfrac{7.50 \text{ g KCl}}{63.75 \text{ g}} \times 100 = 11.8 \text{ \%KCl}$

b. mass of solution = 4.25 + 7.50 + 125 = 136.75 g (137 g)

$\dfrac{4.25 \text{ g NaCl}}{136.75 \text{ g}} \times 100 = 3.11 \text{ \%NaCl}$ $\dfrac{7.50 \text{ g KCl}}{136.75 \text{ g}} \times 100 = 5.48 \text{ \% KCl}$

c. mass of solution = 4.25 + 7.50 + 355 = 366.75 g (367 g)

$\dfrac{4.25 \text{ g NaCl}}{366.75 \text{ g}} \times 100 = 1.16 \text{ \%NaCl}$ $\dfrac{7.50 \text{ g KCl}}{366.75 \text{ g}} \times 100 = 2.04 \text{ \%KCl}$

36. a. $\dfrac{(3.02 \text{ } M)(255 \text{ mL})}{(255 + 375) \text{ mL}} = 1.22 \text{ } M$

b. $\dfrac{(1.51 \text{ \%})(75.1 \text{ g})}{(75.1 + 125) \text{ g}} = 0.567 \text{ \%}$

c. $\dfrac{(12.1 \text{ } M)(6.25 \text{ mL})}{(6.25 + 490.) \text{ mL}} = 0.152 \text{ } M$

38. a. $\dfrac{(41.5 \text{ mL})(0.118 \text{ } M)(1)}{(0.242 \text{ } M)(2)} = 10.1 \text{ mL H}_2\text{SO}_4$

b. $\dfrac{(27.1 \text{ mL})(0.121 \text{ } M)(3)}{(0.242 \text{ } M)(2)} = 20.3 \text{ mL H}_2\text{SO}_4$

Chapter 15   Acids and Bases

2. In the Arrhenius definition, an acid is a substance which produces hydrogen ions (H^+) when dissolved in water, whereas a base is a substance which produces hydroxide ions (OH^-) in aqueous solution. These definitions proved to be too restrictive since the only base permitted was hydroxide ion, and the only solvent permitted was water.

4. A conjugate acid-base pair differ from each other by one proton (one hydrogen ion, H^+). For example, CH_3COOH (acetic acid), differs from its conjugate base, CH_3COO^- (acetate ion), by a single H^+ ion.

    $$CH_3COOH(aq) \rightarrow CH_3COO^-(aq) + H^+(aq)$$

6. When an acid is dissolved in water, the hydronium ion (H_3O^+) is formed. The hydronium ion is the conjugate *acid* of water (H_2O).

8. a. HSO_4^- and SO_4^{2-} represent a conjugate acid-base pair (HSO_4^- is the acid, SO_4^{2-} is the base; they differ by one proton).

    b. HBr and BrO^- are not a conjugate acid-base pair (Br^- is the conjugate base of HBr; BrO^- is the conjugate base of $HBrO$).

    c. $H_2PO_4^-$ and PO_4^{3-} are not a conjugate acid-base pair; they differ by *two* protons ($H_2PO_4^-$ is the conjugate acid of HPO_4^{2-} and also the conjugate base of H_3PO_4; HPO_4^{2-} is the conjugate acid of PO_4^{3-}).

    d. HNO_3 and NO_2^- are not a conjugate acid-base pair; they differ by an oxygen atom as well as a proton (NO_3^- is the conjugate base of HNO_3; NO_2^- is the conjugate base of HNO_2).

10. a. NH_3 (base), H_2O (acid); NH_4^+ (acid), OH^- (base)

    b. PO_4^{3-} (base), H_2O (acid); HPO_4^{2-} (acid), OH^- (base)

    c. $C_2H_3O_2^-$ (base), H_2O (acid); $HC_2H_3O_2$ (acid), OH^- (base)

12. The conjugate *acid* of the species indicated would have *one additional proton*:

    a. HSO_4^-

    b. HPO_4^{2-}

    c. $CH_3NH_3^+$

    d. HF

14. The conjugate *bases* of the species indicated would have *one less proton*:

    a. CO_3^{2-}
    b. HPO_4^{2-}
    c. Cl^-
    d. SO_4^{2-}

16. a. $HClO_4 + H_2O \rightarrow ClO_4^- + H_3O^+$
    b. $HC_2H_3O_2 + H_2O \rightleftarrows C_2H_3O_2^- + H_3O^+$
    c. $HSO_3^- + H_2O \rightleftarrows SO_3^{2-} + OH^-$
    d. $HBr + H_2O \rightarrow Br^- + H_3O^+$

18. To say that an acid is *weak* in aqueous solution means that the acid does not easily transfer protons to water (and does not fully ionize). If an acid does not lose protons easily, then the acid's anion must be a strong attractor of protons (good at holding on to protons).

20. A strong acid is one which loses its protons easily and fully ionizes in water; this means that the acid's conjugate base must be poor at attracting and holding on to protons, and is therefore a relatively weak base. A weak acid is one which resists loss of its protons and does not ionize well in water; this means that the acid's conjugate base attracts and holds onto protons tightly and is a relatively strong base.

22. H_2SO_4 (sulfuric): $H_2SO_4 + H_2O \rightarrow HSO_4^- + H_3O^+$

    HCl (hydrochloric): $HCl + H_2O \rightarrow Cl^- + H_3O^+$

    HNO_3 (nitric): $HNO_3 + H_2O \rightarrow NO_3^- + H_3O^+$

    $HClO_4$ (perchloric): $HClO_4 + H_2O \rightarrow ClO_4^- + H_3O^+$

24. oxyacids: $HClO_4$, HNO_3, H_2SO_4, CH_3COOH, etc.

    non-oxyacids: HCl, HBr, HF, HI, HCN, etc.

26. Bases that are *weak* have relatively strong conjugate acids:

    a. SO_4^{2-} is a moderately weak base; HSO_4^- is a moderately strong acid
    b. Br^- is a very weak base; HBr is a strong acid
    c. CN^- is a fairly strong base; HCN is a weak acid
    d. CH_3COO^- is a fairly strong base; CH_3COOH is a weak acid

28. For example, HCO_3^- can behave as an acid if it reacts with something that more strongly gains protons than does HCO_3^- itself. For example, HCO_3^- would behave as an acid when reacting with hydroxide ion (a much stronger base).

    $HCO_3^-(aq) + OH^-(aq) \rightarrow CO_3^{2-}(aq) + H_2O(l)$

    On the other hand, HCO_3^- would behave as a base when reacted with something that more readily loses protons than does HCO_3^- itself. For example, HCO_3^- would behave as a base when reacting with hydrochloric acid (a much stronger acid).

    $HCO_3^-(aq) + HCl(aq) \rightarrow H_2CO_3(aq) + Cl^-(aq)$

For $H_2PO_4^-$, similar equations can be written:

$$H_2PO_4^-(aq) + OH^-(aq) \rightarrow HPO_4^{2-}(aq) + H_2O(l)$$

$$H_2PO_4^-(aq) + H_3O^+(aq) \rightarrow H_3PO_4(aq) + H_2O(l)$$

30. The hydrogen ion concentration and the hydroxide ion concentration of water are *not* independent of each other: they are related by the equilibrium

$$H_2O(l) \rightleftharpoons H^+(aq) + OH^-(aq)$$

for which $K_w = [H^+][OH^-] = 1.0 \times 10^{-14}$ at 25°C.

If the concentration of one of these ions is increased by addition of a reagent producing H^+ or OH^-, then the concentration of the complementary ion will have to decrease so that the value of K_w will hold true. So if an acid is added to a solution, the concentration of hydroxide ion in the solution will decrease to a lower value. Similarly, if a base is added to a solution, then the concentration of hydrogen ion will have to decrease to a lower value.

32. $K_w = [H^+][OH^-] = 1.0 \times 10^{-14}$ at 25°C

$$[H^+] = \frac{1.0 \times 10^{-14}}{[OH^-]}$$

a. $[H^+] = \dfrac{1.0 \times 10^{-14}}{3.99 \times 10^{-5} \ M} = 2.5 \times 10^{-10} \ M$; solution is basic

b. $[H^+] = \dfrac{1.0 \times 10^{-14}}{2.91 \times 10^{-9} \ M} = 3.4 \times 10^{-6} \ M$; solution is acidic

c. $[H^+] = \dfrac{1.0 \times 10^{-14}}{7.23 \times 10^{-2} \ M} = 1.4 \times 10^{-13} \ M$; solution is basic

d. $[H^+] = \dfrac{1.0 \times 10^{-14}}{9.11 \times 10^{-7} \ M} = 1.1 \times 10^{-8} \ M$; solution is basic

34. $[OH^-] = \dfrac{1.0 \times 10^{-14}}{[H^+]}$

a. $[OH^-] = \dfrac{1.0 \times 10^{-14}}{1.00 \times 10^{-7} \ M} = 1.0 \times 10^{-7} \ M$; solution is neutral

b. $[OH^-] = \dfrac{1.0 \times 10^{-14}}{7.00 \times 10^{-7} \ M} = 1.4 \times 10^{-8} \ M$; solution is acidic

c. $[OH^-] = \dfrac{1.0 \times 10^{-14}}{7.00 \times 10^{-1} \ M} = 1.4 \times 10^{-14} \ M$; solution is acidic

d. $[OH^-] = \dfrac{1.0 \times 10^{-14}}{5.99 \times 10^{-6} \ M} = 1.7 \times 10^{-9} \ M$; solution is acidic

36. a. $[H^+] = 1.04 \times 10^{-8} \ M$ is more basic

   b. $[OH^-] = 4.49 \times 10^{-6} \ M$ is more basic

   c. $[OH^-] = 6.01 \times 10^{-7} \ M$ is more basic

38. household ammonia (pH 12); blood (pH 7-8); milk (pH 6-7); vinegar (pH 3); lemon juice (pH 2-3); stomach acid (pH 2)

40. The pH of a solution is defined as the *negative* of the logarithm of the hydrogen ion concentration, pH = $-\log[H^+]$. Mathematically, the *negative sign* in the definition causes the pH to *decrease* as the hydrogen ion concentration *increases*.

42. pH = $-\log[H^+]$

   a. pH = $-\log[0.0010 \ M] = 3.000$; solution is acidic

   b. pH = $-\log[2.19 \times 10^{-4} \ M] = 3.660$; solution is acidic

   c. pH = $-\log[9.18 \times 10^{-11} \ M] = 10.037$; solution is basic

   d. pH = $-\log[4.71 \times 10^{-7} \ M] = 6.327$; solution is acidic

44. pOH = $-\log[OH^-]$ pH = 14 - pOH

   a. pOH = $-\log[1.00 \times 10^{-7} \ M] = 7.000$

      pH = 14 - 7.000 = 7.000; solution is neutral

   b. pOH = $-\log[4.59 \times 10^{-13} \ M] = 12.338$

      pH = 14 - 12.338 = 1.662; solution is acidic

   c. pOH = $-\log[1.04 \times 10^{-4} \ M] = 3.983$

      pH = 14 - 3.983 = 10.017; solution is basic

   d. pOH = $-\log[7.00 \times 10^{-1} \ M] = 0.155$

      pH = 14 - 0.155 = 13.845; solution is basic

46. pOH = 14 - pH

   a. pOH = 14 - 6.49 = 7.51; solution is acidic

   b. pOH = 14 - 1.93 = 12.07; solution is acidic

   c. pOH = 14 - 11.21 = 2.79; solution is basic

204   Chapter 15   Acids and Bases

    d.   pOH = 14 - 7.00 = 7.00; solution is neutral

48.   a. $[H^+] = 1.00 \times 10^{-7}$ M

$$[OH^-] = \frac{1.0 \times 10^{-14}}{1.00 \times 10^{-7} \, M} = 1.0 \times 10^{-7} \, M$$

$pH = -\log[1.0 \times 10^{-7} \, M] = 7.00$

$pOH = 14 - 7.00 = 7.00$

    b. $[OH^-] = 4.39 \times 10^{-5}$ M

$$[H^+] = \frac{1.0 \times 10^{-14}}{4.39 \times 10^{-5} \, M} = 2.28 \times 10^{-10} \, M = 2.3 \times 10^{-10} \, M$$

$pH = -\log[2.28 \times 10^{-10} \, M] = 9.64$

$pOH = 14 - 9.64 = 4.36$

    c. $[H^+] = 4.29 \times 10^{-11}$ M

$$[OH^-] = \frac{1.0 \times 10^{-14}}{4.29 \times 10^{-11} \, M} = 2.33 \times 10^{-4} \, M = 2.3 \times 10^{-4} \, M$$

$pH = -\log[4.29 \times 10^{-11} \, M] = 10.37$

$pOH = 14 - 10.368 = 3.63$

    d. $[OH^-] = 7.36 \times 10^{-2}$ M

$$[H^+] = \frac{1.0 \times 10^{-14}}{7.36 \times 10^{-2} \, M} = 1.36 \times 10^{-13} \, M = 1.4 \times 10^{-13} \, M$$

$pH = -\log[1.36 \times 10^{-13} \, M] = 12.87$

$pOH = 14 - 12.87 = 1.13$

50. $[H^+] = \{inv\}\{log\}[-pH]$

    a. $[H^+] = \{inv\}\{log\}[-1.04] = 9.2 \times 10^{-2}$ M

    b. $[H^+] = \{inv\}\{log\}[-13.1] = 8 \times 10^{-14}$ M

    c. $[H^+] = \{inv\}\{log\}[-5.99] = 1.0 \times 10^{-6}$ M

    d. $[H^+] = \{inv\}\{log\}[-8.62] = 2.4 \times 10^{-9}$ M

52.   a. $pH = 14 - 3.91 = 10.09$

$[H^+] = \{inv\}\{log\}[-10.09] = 8.1 \times 10^{-11}$ M

    b. $pH = 14 - 12.56 = 1.44$

$[H^+] = \{inv\}\{log\}[-1.44] = 3.6 \times 10^{-2}$ M

c. pH = 14 - 1.15 = 12.85

$[H^+]$ = {inv}{log}[-12.85] = 1.4 × 10^{-13} M

d. pH = 14 - 8.77 = 5.23

$[H^+]$ = {inv}{log}[-5.23] = 5.9 × 10^{-6} M

54. a. pOH = 14 - 5.12 = 8.88

$[H^+]$ = {inv}{log}[-5.12] = 7.6 × 10^{-6} M

$[OH^-]$ = {inv}{log}[-8.88] = 1.3 × 10^{-9} M

b. pH = 14 - 5.12 = 8.88

$[H^+]$ = {inv}{log}[-8.88] = 1.3 × 10^{-9} M

$[OH^-]$ = {inv}{log}[-5.12] = 7.6 × 10^{-6} M

c. pOH = 14 - 7.00 = 7.00

$[H^+]$ = $[OH^-]$ {inv}{log}[-7.00] = 1.0 × 10^{-7} M

d. pH = 14 - 13.00 = 1.00

$[H^+]$ = {inv}{log}[-1.00] = 1.0 × 10^{-1} M

$[OH^-]$ = {inv}{log}[-13.00] = 1.0 × 10^{-13} M

56. The solution contains water molecules, H_3O^+ ions (protons), and NO_3^- ions. Because HNO_3 is a strong acid, which is completely ionized in water, there are no HNO_3 molecules present.

58. a. HCl is a strong acid and completely ionized so

$[H^+]$ = 0.00010 M and pH = 4.00

b. HNO_3 is a strong acid and completely ionized so

$[H^+]$ = 0.0050 M and pH = 2.30

c. $HClO_4$ is a strong acid and completely ionized so

$[H^+]$ = 4.21 × 10^{-5} M and pH = 4.376

d. HNO_3 is a strong acid and completely ionized so

$[H^+]$ = 6.33 × 10^{-3} M and pH = 2.199

206 Chapter 15 Acids and Bases

60. A buffered solution consists of a mixture of a weak acid and its conjugate base; one example of a buffered solution is a mixture of acetic acid (CH_3COOH) and sodium acetate ($NaCH_3COO$).

62. The weak acid component of a buffered solution is capable of reacting with added strong base. For example, using the buffered solution given as an example in Question 60, acetic acid would consume added sodium hydroxide as follows:

$$CH_3COOH(aq) + NaOH(aq) \rightarrow NaCH_3COO(aq) + H_2O(l)$$

Acetic acid *neutralizes* the added NaOH and prevents it from having much effect on the overall pH of the solution.

64. added NaOH: $CH_3COOH + OH^- \rightarrow CH_3COO^- + H_2O$
    added HCl: $CH_3COO^- + H_3O^+ \rightarrow CH_3COOH + H_2O$

    added NaOH: $H_2S + OH^- \rightarrow HS^- + H_2O$
    added HCl: $HS^- + H_3O^+ \rightarrow H_2S + H_2O$

66. a. NaOH is completely ionized, so $[OH^-] = 0.10\ M$

    $pOH = -\log[0.10] = 1.00$

    $pH = 14 - 1.00 = 13.00$

    b. KOH is completely ionized, so $[OH^-] = 2.0 \times 10^{-4}\ M$

    $pOH = -\log[2.0 \times 10^{-4}] = 3.70$

    $pH = 14 - 3.70 = 10.30$

    c. CsOH is completely ionized, so $[OH^-] = 6.2 \times 10^{-3}\ M$

    $pOH = -\log[6.2 \times 10^{-3}] = 2.21$

    $pH = 14 - 2.21 = 11.79$

    d. NaOH is completely ionized, so $[OH^-] = 0.0001\ M$

    $pOH = -\log[0.0001] = 4.0$

    $pH = 14 - 4.0 = 10.0$

68. a, b, and d represent basic solutions; c represents an *acidic* solution because there is less hydroxide ion than hydrogen ion.

70. a, c, and e represent strong acids; b and d are typical *weak* acids.

72. Ordinarily in calculating the pH of strong acid solutions, the major contribution to the concentration of hydrogen ion present is from the

dissolved strong acid; we ordinarily neglect the small amount of hydrogen ion present in such solutions due to the ionization of water. With 1.0×10^{-7} M HCl solution, however, the amount of hydrogen ion present due to the ionization of *water* is *comparable* to that present due to the addition of *acid* (HCl) and must be considered in the calculation of pH.

74. accepts

76. base

78. carboxyl (-COOH)

80. 1.0×10^{-14}

82. higher

84. pH

86. weak acid

88.
    a. H_2O and OH^- represent a conjugate acid-base pair (H_2O is the acid, having one more proton than the base, OH^-).

    b. H_2SO_4 and SO_4^{2-} are *not* a conjugate acid-base pair (they differ by *two* protons). The conjugate base of H_2SO_4 is HSO_4^-; the conjugate acid of SO_4^{2-} is also HSO_4^-.

    c. H_3PO_4 and $H_2PO_4^-$ represent a conjugate acid-base pair (H_3PO_4 is the acid, having one more proton than the base $H_2PO_4^-$).

    d. $HC_2H_3O_2$ and $C_2H_3O_2^-$ represent a conjugate acid-base pair ($HC_2H_3O_2$ is the acid, having one more proton than the base $C_2H_3O_2^-$).

90. The conjugate *acid* of the species indicated would have *one additional proton*:

    a. NH_4^+
    b. NH_3
    c. H_3O^+
    d. H_2O

92. When an acid ionizes in water, a proton is released to the water as an H_3O^+ ion:

    a. $CH_3CH_2COOH + H_2O \rightleftharpoons CH_3CH_2COO^- + H_3O^+$
    b. $NH_4^+ + H_2O \rightleftharpoons NH_3 + H_3O^+$

208 Chapter 15 Acids and Bases

    c. $H_2SO_4 + H_2O \rightarrow HSO_4^- + H_3O^+$

    d. $H_3PO_4 + H_2O \rightleftarrows H_2PO_4^- + H_3O^+$

94. $K_w = [H^+][OH^-] = 1.0 \times 10^{-14}$ at 25°C

$$[H^+] = \frac{1.0 \times 10^{-14}}{[OH^-]}$$

  a. $[H^+] = \dfrac{1.0 \times 10^{-14}}{4.22 \times 10^{-3} \, M} = 2.4 \times 10^{-12} \, M$; solution is basic

  b. $[H^+] = \dfrac{1.0 \times 10^{-14}}{1.01 \times 10^{-13} \, M} = 9.9 \times 10^{-2} \, M$; solution is acidic

  c. $[H^+] = \dfrac{1.0 \times 10^{-14}}{3.05 \times 10^{-7} \, M} = 3.3 \times 10^{-8} \, M$; solution is basic

  d. $[H^+] = \dfrac{1.0 \times 10^{-14}}{6.02 \times 10^{-6} \, M} = 1.7 \times 10^{-9} \, M$; solution is basic

96.   a. $[OH^-] = 0.0000032 \, M$ is more basic

    b. $[OH^-] = 1.54 \times 10^{-8} \, M$ is more basic

    c. $[OH^-] = 4.02 \times 10^{-7} \, M$ is more basic

98. pOH = $-\log[OH^-]$ pH = 14 − pOH

  a. pOH = $-\log[1.4 \times 10^{-6} \, M] = 5.85$

     pH = 14 − 5.85 = 8.15; solution is basic

  b. pOH = $-\log[9.35 \times 10^{-9} \, M] = 8.029 = 8.03$

     pH = 14 − 8.029 = 5.971 = 5.97; solution is acidic

  c. pOH = $-\log[2.21 \times 10^{-1} \, M] = 0.656 = 0.66$

     pH = 14 − 0.656 = 13.344 = 13.34; solution is basic

  d. pOH = $-\log[7.98 \times 10^{-12} \, M] = 11.098 = 11.10$

     pH = 14 − 11.098 = 2.902 = 2.90; solution is acidic

100. a. $[OH^-] = \dfrac{1.0 \times 10^{-14}}{5.72 \times 10^{-4} \, M} = 1.75 \times 10^{-11} \, M = 1.8 \times 10^{-11} \, M$

     pOH = $-\log[1.75 \times 10^{-11} \, M] = 10.76$

     pH = 14 − 10.76 = 3.24

  b. $[H^+] = \dfrac{1.0 \times 10^{-14}}{8.91 \times 10^{-5} \, M} = 1.12 \times 10^{-10} \, M = 1.1 \times 10^{-10} \, M$

     pH = $-\log[1.12 \times 10^{-10} \, M] = 9.95$

     pOH = 14 − 9.95 = 4.05

c. $[OH^-] = \dfrac{1.0 \times 10^{-14}}{2.87 \times 10^{-12} \, M} = 3.48 \times 10^{-3} \, M = 3.5 \times 10^{-3} \, M$

$pOH = -\log[3.48 \times 10^{-3} \, M] = 2.46$

$pH = 14 - 2.46 = 11.54$

d. $[H^+] = \dfrac{1.0 \times 10^{-14}}{7.22 \times 10^{-8} \, M} = 1.39 \times 10^{-7} \, M = 1.4 \times 10^{-7} \, M$

$pH = -\log[1.39 \times 10^{-7} \, M] = 6.86$

$pOH = 14 - 6.86 = 7.14$

102. $pH = 14 - pOH \qquad [H^+] = \{inv\}\{log\}[-pH]$

a. $[H^+] = \{inv\}\{log\}[-5.41] = 3.9 \times 10^{-6} \, M$

b. $pH = 14 - 12.04 = 1.96$

$[H^+] = \{inv\}\{log\}[-1.96] = 1.1 \times 10^{-2} \, M$

c. $[H^+] = \{inv\}\{log\}[-11.91] = 1.2 \times 10^{-12} \, M$

d. $pH = 14 - 3.89 = 10.11$

$[H^+] = \{inv\}\{log\}[-10.11] = 7.8 \times 10^{-11} \, M$

104. a. $HClO_4$ is a strong acid and completely ionized so

$[H^+] = 1.4 \times 10^{-3} \, M$ and $pH = 2.85$

b. HCl is a strong acid and completely ionized so

$[H^+] = 3.0 \times 10^{-5} \, M$ and $pH = 4.52$

c. HNO_3 is a strong acid and completely ionized so

$[H^+] = 5.0 \times 10^{-2} \, M$ and $pH = 1.30$

d. HCl is a strong acid and completely ionized so

$[H^+] = 0.0010 \, M$ and $pH = 3.00$

Chapter 16   Equilibrium

2. The nitrogen-nitrogen triple bond in N_2 and the three hydrogen-hydrogen bonds (in the three H_2 molecules) must be broken. Six nitrogen-hydrogen bonds must form (in the two ammonia molecules).

4. The activation energy is the minimum energy two colliding molecules must possess in order for the collision to result in reaction. If molecules do not possess energies equal to or greater than E_a, a collision between these molecules will not result in a reaction.

6. Living cells contain biological catalysts called *enzymes*. Such enzymes are necessary to speed up the complicated biochemical processes that must occur in cells. Such processes would be too slow to sustain life at room temperature if such catalysts were not present.

8. A state of equilibrium is attained when two opposing processes are exactly balanced. The development of a vapor pressure above a liquid in a closed container is an example of a physical equilibrium. Any chemical reaction which appears to "stop" before completion serves as an example of a chemical equilibrium.

10. Chemical equilibrium occurs when two *opposing* chemical reactions reach the *same speed* in a closed system. When a state of chemical equilibrium has been reached, the concentrations of reactants and products present in the system remain *constant* with time, and the reaction appears to "stop." A chemical reaction that reaches a state of equilibrium is indicated by using a double arrow ($\rightleftarrows$). The points of the double arrow point in opposite directions, indicating that two opposite processes are going on.

12. Although we recognize a state of chemical equilibrium by the fact that the concentrations of reactants and products no longer change with time, the lack of change results from the fact that two *opposing* processes are going on at the same time with the same rate (not because the reaction has truly "stopped"). Further reaction in the forward direction is canceled out by an equal extent of reaction in the reverse direction. The reaction is still proceeding, but the opposite reaction is also proceeding at the same rate.

14. The equilibrium constant is a *ratio* of the concentration of products to the concentration of reactants, with all concentrations measured at equilibrium. Depending on how much reactant a particular experiment was begun with, there may be different absolute amounts of reactants and products present at equilibrium, but the *ratio* will always be the same for a given reaction at a given temperature. For example, the ratios (4/2) and (6/3) are different absolutely in terms of the numbers involved, but each of these ratios has the *value* of 2.

16. a. $K = \dfrac{[NCl_3]^2}{[N_2][Cl_2]^3}$

b. $\quad K = \dfrac{[HI]^2}{[H_2]^2[I_2]^2}$

c. $\quad K = \dfrac{[N_2H_4]}{[N_2][H_2]^2}$

18. a. $\quad K = \dfrac{[CH_3OH]}{[CO][H_2]^2}$

b. $\quad K = \dfrac{[NO]^2[O_2]}{[NO_2]^2}$

c. $\quad K = \dfrac{[PBr_3]^4}{[P_4][Br_2]^6}$

20. $N_2(g) + 3H_2(g) \rightleftarrows 2NH_3(g)$

$$K = \dfrac{[NH_3]^2}{[N_2][H_2]^3} = \dfrac{[0.34\ M]^2}{[4.9 \times 10^{-4}\ M][2.1 \times 10^{-3}\ M]^3} = 2.5 \times 10^{10}$$

22. $N_2(g) + 3Cl_2(g) \rightleftarrows 2NCl_3(g)$

$$K = \dfrac{[NCl_3]^2}{[N_2][Cl_2]^3} = \dfrac{[0.141\ M]^2}{[0.000104\ M][0.000201\ M]^3} = 2.35 \times 10^{13}$$

24. Equilibrium constants represent ratios of the *concentrations* of products and reactants present at the point of equilibrium. The *concentration* of a pure solid or of a pure liquid is constant and is determined by the density of the solid or liquid. For example, suppose you had a liter of water. Within that liter of water are 55.5 mol of water (the number of moles of water that is contained in one liter of water *does not vary*).

26. a. $\quad K = [H_2O(g)][CO_2(g)]$

b. $\quad K = [CO_2]$

c. $\quad K = \dfrac{1}{[O_2]^3}$

28. a. $\quad K = [Br_2]^3[N_2]$

b. $\quad K = \dfrac{[H_2O]}{[H_2]}$

c. $\quad K = [CO_2]$

30. When an additional amount of one of the reactants is added to an equilibrium system, the system shifts to the right and adjusts so as to use up some of the added reactant. This results in a net *increase* in the amount of product, compared to the equilibrium system before the additional reactant was added. The numerical *value* of the equilibrium

Chapter 16   Equilibrium

constant does *not* change when a reactant is added: the concentrations of all reactants and products adjust until the correct value of K is once again achieved.

32. If heat is applied to an endothermic reaction (i.e., the temperature is raised), the equilibrium is shifted to the right. More product will be present at equilibrium than if the temperature had not been increased. The value of K increases.

34.
   a. shifts right (system reacts to get rid of excess SO_2)
   b. shifts right (system reacts to replace missing SO_3)
   c. no change (catalysts do not affect the position of equilibrium)

36.
   a. no effect (UO_2 is a solid)
   b. no effect (Xe is not involved in the reaction)
   c. shifts to left (if HF attacks glass, it is removed from system, causing reaction to replace the lost HF)
   d. shifts to right
   e. shifts to left (4 mol gas versus 3 mol gas)

38. A increase in temperature will tend to increase the yield of product. Heat is a reactant for the reaction; adding heat favors the forward reaction.

40. For an *endo*thermic reaction, an increase in temperature will shift the position of equilibrium to the right (toward products).

42. Heat is a product of the reaction. Removing heat will tend to favor the forward reaction. The reaction is *exo*thermic and should be performed at as low a temperature as possible (consistent with the molecules still having sufficient energy to react).

44. A small equilibrium constant means that the concentration of products is small, compared to the concentration of reactants. The position of equilibrium lies far to the left. Reactions with very small equilibrium constants are generally not very useful as a source of the products, unless Le Châtelier's principle can be applied to shift the position of equilibrium to the point where a sufficient amount of product can be isolated.

46. $K = \dfrac{[NCl_3]^2}{[N_2][Cl_2]^3} = \dfrac{[1.9 \times 10^{-1}]^2}{[1.4 \times 10^{-3}\ M][4.3 \times 10^{-4}\ M]^3} = 3.2 \times 10^{11}$

48. $K = [CO_2] = [2.1 \times 10^{-3}\ M] = 2.1 \times 10^{-3}$

50. $K = \dfrac{[H_2]^2[O_2]}{[H_2O]^2}$

$$2.4 \times 10^{-3} = \frac{[1.9 \times 10^{-2}]^2 [O_2]}{[1.1 \times 10^{-1}]^2}$$

$$[O_2] = \frac{(2.4 \times 10^{-3})(1.1 \times 10^{-1})^2}{(1.9 \times 10^{-2})^2} = 8.0 \times 10^{-2}\ M$$

52. $K = \dfrac{[NO_2]^2}{[N_2O_4]} = 8.1 \times 10^{-3}$

$$8.1 \times 10^{-3} = \frac{[0.0021\ M]^2}{[N_2O_4]}$$

$[N_2O_4] = 5.4 \times 10^{-4}\ M$

54. solubility product, K_{sp}

56. Stirring or grinding the solute increases the speed with which the solute dissolves, but the ultimate *amount* of solute that dissolves is fixed by the equilibrium constant for the dissolving process, K_{sp}.

58. a. $PbBr_2(s) \rightleftharpoons Pb^{2+}(aq) + 2Br^-(aq)$

$K_{sp} = [Pb^{2+}][Br^-]^2$

b. $Ag_2S(s) \rightleftharpoons 2Ag^+(aq) + S^{2-}(aq)$

$K_{sp} = [Ag^+]^2[S^{2-}]$

c. $PbCO_3(s) \rightleftharpoons Pb^{2+}(aq) + CO_3^{2-}(aq)$

$K_{sp} = [Pb^{2+}][CO_3^{2-}]$

d. $Sr_3(PO_4)_2(s) \rightleftharpoons 3Sr^{2+}(aq) + 2PO_4^{3-}(aq)$

$K_{sp} = [Sr^{2+}]^3[PO_4^{3-}]^2$

60. $ZnCO_3(s) \rightleftharpoons Zn^{2+}(aq) + CO_3^{2-}(aq)$

$K_{sp} = [Zn^{2+}][CO_3^{2-}]$

If 1.7×10^{-5} mol of $ZnCO_3$ dissolve per liter, then the concentrations of the two ions produced will each also be $1.7 \times 10^{-5}\ M$.

$K_{sp} = [1.7 \times 10^{-5}\ M][1.7 \times 10^{-5}\ M] = 2.9 \times 10^{-10}$.

62. $CuCrO_4(s) \rightleftharpoons Cu^{2+}(aq) + CrO_4^{2-}(aq)$

$K_{sp} = [Cu^{2+}][CrO_4^{2-}]$

Molar mass of $CuCrO_4$ = 179.55 g

214 Chapter 16 Equilibrium

$$1.1 \times 10^{-5} \text{ g CuCrO}_4 \times \frac{1 \text{ mol}}{179.55 \text{ g}} = 6.1 \times 10^{-8} \text{ mol}$$

If 6.1×10^{-8} mol of $CuCrO_4$ dissolve per liter, then the concentrations of $Cu^{2+}(aq)$ and $CrO_4^{2-}(aq)$ are also each 6.1×10^{-8} M.

$$K_{sp} = (6.1 \times 10^{-8} M)(6.1 \times 10^{-8} M) = 3.8 \times 10^{-15}$$

64. $CuCrO_4(s) \rightleftharpoons Cu^{2+}(aq) + CrO_4^{2-}(aq)$

$$K_{sp} = [Cu^{2+}][CrO_4^{2-}] = 3.8 \times 10^{-6}$$

Let x represent the number of moles of $CuCrO_4(s)$ that dissolve per liter, then $[Cu^{2+}] = x$ and $[CrO_4^{2-}] = x$ also from the stoichiometry of the reaction; then

$$K_{sp} = [x][x] = x^2 = 3.8 \times 10^{-6}$$

$$x = [CuCrO_4] = 1.9 \times 10^{-3} M$$

66. $PbCrO_4(s) \rightleftharpoons Pb^{2+}(aq) + CrO_4^{2-}(aq)$

$$K_{sp} = [Pb^{2+}][CrO_4^{2-}] = 2.8 \times 10^{-13}$$

Let x represent the number of moles of $PbCrO_4(s)$ that dissolve per liter. Then $[Pb^{2+}] = x$ and $[CrO_4^{2-}] = x$ also from the stoichiometry of the reaction; then

$$K_{sp} = [x][x] = x^2 = 2.8 \times 10^{-13}$$

$$x = [PbCrO_4] = 5.3 \times 10^{-7} M.$$

Molar mass of $PbCrO_4 = 323.2$ g

5.3×10^{-7} mol/L $\times 323.2$ g/mol $= 1.7 \times 10^{-4}$ g/L

68. $PbCl_2(s) \rightleftharpoons Pb^{2+}(aq) + 2Cl^-(aq)$

$$K_{sp} = [Pb^{2+}][Cl^-]^2$$

If $PbCl_2$ dissolves to the extent of $3.6 \times 10^{-2} M$, then

$[Pb^{2+}] = 3.6 \times 10^{-2} M$ and

$[Cl^-] = 2 \times (3.6 \times 10^{-2}) = 7.2 \times 10^{-2} M$

$$K_{sp} = (3.6 \times 10^{-2} M)(7.2 \times 10^{-2} M)^2 = 1.9 \times 10^{-4}$$

molar mass $PbCl_2 = 278.1$ g

$$3.6 \times 10^{-2} \frac{\text{mol}}{1 \text{ L}} \times \frac{278.1 \text{ g}}{1 \text{ mol}} = 10. \text{ g/L}$$

70. $Fe(OH)_3(s) \rightleftarrows Fe^{3+}(aq) + 3OH^-(aq)$

    $K_{sp} = [Fe^{3+}][OH^-]^3 = 4 \times 10^{-38}$

    Let x represent the number of moles of $Fe(OH)_3$ that dissolve per liter; then $[Fe^{3+}] = x$.

    The amount of hydroxide ion that would be produced by the dissolving of the $Fe(OH)_3$ would then be $3x$, but pure water itself contains hydroxide ion at the concentration of 1.0×10^{-7} M (see Chapter 17). The total concentration of hydroxide ion is then $[OH^-] = (3x + 1.0 \times 10^{-7})$. Since x must be a very small number [since $Fe(OH)_3$ is not very soluble], we can save ourselves a lot of arithmetic if we use the approximation that $(3x + 1.0 \times 10^{-7} M) = 1.0 \times 10^{-7}$

    $K_{sp} = [x][1.0 \times 10^{-7}]^3 = 4 \times 10^{-38}$

    $x = 4 \times 10^{-17}$ M

    molar mass $Fe(OH)_3 = 106.9$ g

    $$\frac{4 \times 10^{-17} \text{ mol}}{1.00 \text{ L}} \times \frac{106.9 \text{ g}}{1 \text{ mol}} = 4 \times 10^{-15} \text{ g/L}$$

72. An increase in temperature increases the fraction of molecules that possess sufficient energy for a collision to result in a reaction.

74. catalyst

76. constant

78. When we say that a chemical equilibrium is *dynamic*, we are recognizing the fact that even though the reaction has appeared macroscopically to have stopped, on a microscopic basis the forward and reverse reactions are still taking place, at the same speed.

80. heterogeneous

82. position

84. Heat is considered a *product* of an exothermic process. Adding a product to a system in equilibrium causes the reverse reaction to occur (producing additional reactants).

86. An equilibrium reaction may come to many *positions* of equilibrium, but at each possible position of equilibrium, the numerical value of the equilibrium constant is fulfilled. If different amounts of reactant are taken in different experiments, the *absolute amounts* of reactant and

216   Chapter 16   Equilibrium

product present at the point of equilibrium reached will differ from one experiment to another, but the *ratio* that defines the equilibrium constant will be the same.

88. $PCl_5(g) \rightleftharpoons PCl_3(g) + Cl_2(g)$

$$K = \frac{[PCl_3][Cl_2]}{[PCl_5]} = 4.5 \times 10^{-3}$$

The concentration of PCl_5 is to be twice the concentration of PCl_3:

$[PCl_5] = 2 \times [PCl_3]$

$$K = \frac{[PCl_3][Cl_2]}{2 \times [PCl_3]} = 4.5 \times 10^{-3}$$

$$K = \frac{[Cl_2]}{2} = 4.5 \times 10^{-3} \quad \text{and} \quad [Cl_2] = 9.0 \times 10^{-3} \, M$$

90. Since all of the metal carbonates indicated have the metal ion in the +2 oxidation state, we can illustrate the calculations for a general metal carbonate, MCO_3:

$$MCO_3(s) \rightleftharpoons M^{2+}(aq) + CO_3^{2-}(aq) \qquad K_{sp} = [M^{2+}(aq)][CO_3^{2-}(aq)]$$

If we then let x represent the number of moles of MCO_3 that dissolve per liter, then $[M^{2+}(aq)] = x$ and $[CO_3^{2-}(aq)] = x$ also since the reaction is of 1:1 stoichiometry. Therefore,

$K_{sp} = [M^{2+}(aq)][CO_3^{2-}(aq)] = x^2$ for each salt. Solving for x gives the following results.

$[BaCO_3] = x = 7.1 \times 10^{-5} \, M$

$[CdCO_3] = x = 2.3 \times 10^{-6} \, M$

$[CaCO_3] = x = 5.3 \times 10^{-5} \, M$

$[CoCO_3] = x = 3.9 \times 10^{-7} \, M$

92. Although a small solubility product generally implies a small solubility, comparisons of solubility based directly on K_{sp} values are only valid if the salts produce the same numbers of positive and negative ions per formula when they dissolve. For example, one can compare the solubilities of $AgCl(s)$ and $NiS(s)$ directly using K_{sp}, since each salt produces one positive and one negative ion per formula when dissolved. One could not directly compare $AgCl(s)$ with a salt such as $Ca_3(PO_4)_2$, however.

94. At higher temperatures, the average kinetic energy of the reactant molecules is larger. At higher temperatures, the probability that a collision between molecules will be energetic enough for reaction to take place is larger. On a molecular basis, a higher temperature means a given molecule will be moving faster.

96. a. $$K = \frac{[HBr]^2}{[H_2][Br_2]}$$

b. $$K = \frac{[H_2S]^2}{[H_2]^2[S_2]}$$

c. $$K = \frac{[HCN]^2}{[H_2][C_2N_2]}$$

98. $$K = \frac{[Br]^2}{[Br_2]} = \frac{(0.034\ M)^2}{(0.97\ M)} = 1.2 \times 10^{-3}$$

100. a. $$K = \frac{1}{[O_2]^3}$$

b. $$K = \frac{1}{[NH_3][HCl]}$$

c. $$K = \frac{1}{[O_2]}$$

102. An *exo*thermic reaction is one which liberates heat energy. Increasing the temperature (adding heat) for such a reaction is fighting against the reaction's own tendency to liberate heat. The net effect of raising the temperature will be a shift to the left to decrease the amount of product. If it is desired to increase the amount of product in an exothermic reaction, heat must be *removed* from the system. Changing the temperature *does* change the numerical value of the equilibrium constant for a reaction.

104. The reaction is *exo*thermic as written. An increase in temperature (addition of heat) will shift the reaction to the left (toward reactants).

106. $$K = \frac{[NH_3]^2}{[N_2][H_2]^3}$$

$$1.3 \times 10^{-2} = \frac{[NH_3]^2}{(0.10\ M)(0.10\ M)^3}$$

$[NH_3]^2 = 1.3 \times 10^{-2} \times (0.10) \times (0.10)^3 = 1.3 \times 10^{-6}$

$[NH_3] = 1.1 \times 10^{-3}\ M$

108. a. $Cu(OH)_2(s) \rightleftarrows Cu^{2+}(aq) + 2OH^-(aq)$

$K_{sp} = [Cu^{2+}][OH^-]^2$

b. $Cr(OH)_3(s) \rightleftarrows Cr^{3+}(aq) + 3OH^-(aq)$

$K_{sp} = [Cr^{3+}][OH^-]^3$

218   Chapter 16   Equilibrium

    c. $Ba(OH)_2(s) \rightleftarrows Ba^{2+}(aq) + 2OH^-(aq)$

           $K_{sp} = [Ba^{2+}][OH^-]^2$

    d. $Sn(OH)_2(s) \rightleftarrows Sn^{2+}(aq) + 2OH^-(aq)$

           $K_{sp} = [Sn^{2+}][OH^-]^2$

110. molar mass AgCl = 143.4 g

$$9.0 \times 10^{-4} \text{ g AgCl/L} \times \frac{1 \text{ mol AgCl}}{143.4 \text{ g AgCl}} = 6.28 \times 10^{-6} \text{ mol AgCl/L}$$

$AgCl(s) \rightleftarrows Ag^+(aq) + Cl^-(aq)$

$K_{sp} = [Ag^+][Cl^-] = (6.28 \times 10^{-6} \, M)(6.28 \times 10^{-6} \, M) = 3.9 \times 10^{-11}$

112. molar mass Ni(OH)$_2$ = 92.71 g

$$\frac{0.14 \text{ g Ni(OH)}_2}{1.00 \text{ L}} \times \frac{1 \text{ mol Ni(OH)}_2}{92.71 \text{ g Ni(OH)}_2} = 1.510 \times 10^{-3} \, M$$

$Ni(OH)_2(s) \rightleftarrows Ni^{2+}(aq) + 2OH^-(aq)$

$K_{sp} = [Ni^{2+}][OH^-]^2$

If 1.510×10^{-3} M of Ni(OH)$_2$ dissolves, then

$[Ni^{2+}] = 1.510 \times 10^{-3}$ M and

$[OH^-] = 2 \times (1.510 \times 10^{-3} \, M) = 3.020 \times 10^{-3} \, M$

$K_{sp} = (1.510 \times 10^{-3} \, M)(3.020 \times 10^{-3} \, M)^2 = 1.4 \times 10^{-8}$

Cumulative Review: Chapters 15 and 16

2. A conjugate acid-base pair consists of two species related to each other by the donating or accepting of a single proton, H^+. An acid has one more H^+ than its conjugate base; a base has one less H^+ than its conjugate acid.

   Brönsted-Lowry acids:

   $$HCl(aq) + H_2O(l) \rightarrow Cl^-(aq) + H_3O^+(aq)$$

   $$H_2SO_4(aq) + H_2O(l) \rightarrow HSO_4^-(aq) + H_3O^+(aq)$$

   $$H_3PO_4(aq) + H_2O(l) \rightleftharpoons H_2PO_4^-(aq) + H_3O^+(aq)$$

   $$NH_4^+(aq) + H_2O(l) \rightleftharpoons NH_3(aq) + H_3O^+(aq)$$

   Brönsted-Lowry bases:

   $$NH_3(aq) + H_2O(l) \rightleftharpoons NH_4^+(aq) + OH^-(aq)$$

   $$HCO_3^-(aq) + H_2O(l) \rightleftharpoons H_2CO_3(aq) + OH^-(aq)$$

   $$NH_2^-(aq) + H_2O(l) \rightarrow NH_3(aq) + OH^-(aq)$$

   $$H_2PO_4^-(aq) + H_2O(l) \rightleftharpoons H_3PO_4(aq) + OH^-(aq)$$

4. The strength of an acid is a direct result of the position of the acid's ionization equilibrium. We call acids whose ionization equilibrium positions lie far to the right strong acids, and we call those acids whose equilibrium positions lie only slightly to the right weak acids. For example, HCl, HNO_3, and $HClO_4$ are all strong acids, which means that they are completely ionized in aqueous solution (the position of equilibrium is very far to the right):

   $$HCl(aq) + H_2O(l) \rightarrow Cl^-(aq) + H_3O^+(aq)$$

   $$HNO_3(aq) + H_2O(l) \rightarrow NO_3^-(aq) + H_3O^+(aq)$$

   $$HClO_4(aq) + H_2O(l) \rightarrow ClO_4^-(aq) + H_3O^+(aq)$$

   Since these are very strong acids, we know that their anions (Cl^-, NO_3^-, ClO_4^-) must be very weak bases, and that solutions of the sodium salts of these anions would *not* be appreciably basic. Since these acids have a strong tendency to lose protons, there is very little tendency for the anions (bases) to gain protons.

6. The pH of a solution is defined as the negative of the base 10 logarithm of the hydrogen ion concentration in the solution; that is

   $$pH = -\log_{10}[H^+]$$

   for a solution. Since in pure water, the amount of $H^+(aq)$ ion present is equal to the amount of $OH^-(aq)$ ion, we say that pure water is *neutral*. Since $[H^+] = 1.0 \times 10^{-7}$ M in pure water, this means that the pH of pure water is $-\log[1.0 \times 10^{-7} M] = 7.00$. Solutions in which the hydrogen ion concentration is greater than $1.0 \times 10^{-7} M$ (pH < 7.00) are *acidic*; solutions in which the hydrogen ion concentration is less than $1.0 \times 10^{-7} M$ (pH > 7.00) are *basic*. Since the pH scale is logarithmic, when the pH changes by one unit, this corresponds to a change in the hydrogen ion concentration by a factor of *ten*.

In some instances, it may be more convenient to speak directly about the hydroxide ion concentration present in a solution, and so an analogous logarithmic expression is defined for the hydroxide ion concentration:

$$pOH = -\log_{10}[OH^-]$$

The concentrations of hydrogen ion and hydroxide ion in water (and in aqueous solutions) are *not* independent of one another, but rather are related by the dissociation equilibrium constant for water,

$$K_w = [H^+][OH^-] = 1.0 \times 10^{-14} \text{ at } 25°C.$$

From this constant it is obvious that pH + pOH = 14.00 for water (or an aqueous solution) at 25°C.

8. Chemists envision that a reaction can only take place between molecules if the molecules physically *collide* with each other. Furthermore, when molecules collide, the molecules must collide with enough force for the reaction to be successful (there must be enough energy to break bonds in the reactants), and the colliding molecules must be positioned with the correct relative orientation for the products (or intermediates) to form. Reactions tend to be faster if higher concentrations are used for the reaction, because, if there are more molecules present per unit volume there will be more collisions between molecules in a given time period. Reactions are faster at higher temperatures because at higher temperatures the reactant molecules have a higher average kinetic energy, and the number of molecules that will collide with sufficient force to break bonds increases.

10. Chemists define equilibrium as the exact balancing of two exactly opposing processes. When a chemical reaction is begun by combining pure reactants, the only process possible initially is

$$\text{reactants} \rightarrow \text{products}$$

However, for many reactions, as the concentration of product molecules increases, it becomes more and more likely that product molecules will collide and react with each other

$$\text{products} \rightarrow \text{reactants}$$

giving back molecules of the original reactants. At some point in the process the rates of the forward and reverse reactions become equal, and the system attains chemical equilibrium. To an outside observer, the system appears to have stopped reacting. On a microscopic basis, though, both the forward and reverse processes are still going on: every time additional molecules of the product form, however, somewhere else in the system molecules of product react to give back molecules of reactant.

Once the point is reached that product molecules are reacting at the same speed at which they are forming, there is no further net change in concentration. A graph showing how the rates of the forward and reverse reactions change with time is given in the text as Figure 16.8. At the start of the reaction, the rate of the forward reaction is at its maximum, while the rate of the reverse reaction is zero. As the reaction proceeds, the rate of the forward reaction gradually decreases as the concentration of reactants decreases, whereas the rate of the reverse reaction increases as the concentration of products increases. Once the two rates have become equal, the reaction has reached a state of equilibrium.

12. The equilibrium constant for a reaction is a *ratio* of the concentration of products present at the point of equilibrium to the concentration of reactants still present. A *ratio* means that we have one number divided by another number (for example, the density of a substance is the ratio of a substance's mass to its volume). Since the equilibrium constant is a ratio, there are an infinite number of sets of data which can give the same ratio: for example, the ratios 8/4, 6/3, 100/50 all have the same value, 2. The actual concentrations of products and reactants will differ from one experiment to another involving a particular chemical reaction, but the ratio of the amount of product to reactant at equilibrium should be the same for each experiment.

Consider this simple example: suppose we have a reaction for which $K = 4$, and we begin this reaction with 100 reactant molecules. At the point of equilibrium, there should be 80 molecules of product and 20 molecules of reactant remaining (80/20 = 4). Suppose we perform another experiment involving the same reaction, only this time we begin the experiment with 500 molecules of reactant. This time, at the point of equilibrium, there will be 400 molecules of product present and 100 molecules of reactant remaining (400/100 = 4). Since we began the two experiments with different numbers of of reactant molecules, it's not troubling that there are different absolute numbers of product and reactant molecules present at equilibrium: however, the ratio, K, is the same for both experiments. We say that these two experiments represent two different positions of equilibrium: an equilibrium position corresponds to a particular set of experimental equilibrium concentrations which fulfill the value of the equilibrium constant. Any experiment that is performed with a different amount of starting material will come to its own unique equilibrium position, but the equilibrium constant ratio, K, will be the same for a given reaction regardless of the starting amounts taken.

14. Your paraphrase of Le Châtelier's principle should go something like this: "when you make any change to a system in equilibrium, this throws the system temporarily out of equilibrium, and the system responds by reacting in whichever direction will be able to reach a new position of equilibrium". There are various changes that can be made to a system in equilibrium. Here are examples of some of them.

a. the concentration of one of the reactants is increased.

Consider the reaction: $2SO_2(g) + O_2(g) \rightleftharpoons 2SO_3(g)$

Suppose the reactants have already reacted and a position of equilibrium has been reached which fulfills the value of K for the reaction. At this point there will be present particular amounts of each reactant and of the product. Suppose then 1 additional mole of O_2 is added to the system from outside. At the instant the additional O_2 is added, the system will not be in equilibrium: there will be too much O_2 present in the system to be compatible with the amounts of SO_2 and SO_3 present. The system will respond by reacting to get rid of some of the excess O_2 until the value of the ratio K is again fulfilled. If the system reacts to get rid of the excess of O_2, additional product SO_3 will form. The net result is more SO_3 produced than if the the change had not been made.

b. the concentration of one of the products is decreased by selectively removing it from the system

Consider the reaction: $CH_3COOH + CH_3OH \rightleftharpoons H_2O + CH_3COOCH_3$

This reaction is typical of many reactions involving organic chemical substances, in which two organic molecules react to form a larger molecule, with a molecule of water split out during the combination. This type of reaction on its own tends to come to equilibrium with only part of the starting materials being converted to the desired organic product (which effectively would leave the experimenter with a mixture of materials). A technique which is used by organic chemists to increase the effective yield of the desired organic product is to *separate* the two products (if the products are separated, they cannot react to give back the reactants). One method used is to add a drying agent to the mixture: such a drying agent chemically or physically absorbs the water from the system, removing it from the equilibrium. If the water is removed, the reverse reaction cannot take place, and the reaction proceeds to a greater extent in the forward direction than if the drying agent had not been added. In other situations, an experimenter may separate the products of the reaction by distillation (if the boiling points make this possible): again, if the products have been separated, then the reverse reaction will not be possible, and the forward reaction will occur to a greater extent.

c. the reaction system is compressed to a smaller volume

Consider the example: $3H_2(g) + N_2(g) \rightleftharpoons 2NH_3(g)$

For equilibria involving gases, when the volume of the reaction system is compressed suddenly, the pressure in the system increases. However, if the reacting system can relieve some of this increased pressure by reacting, it will do so. This will happen by the reaction occurring in whichever direction will give the smaller number of moles of gas (if the number of moles of gas is decreased in a particular volume, the pressure will decrease).

For the reaction above, there are two moles of the gas on the right side of the equation, but there is a total of four moles on the left side. If this system at equilibrium were to be suddenly compressed to a smaller volume, the reaction would proceed further to the right (in favor of more ammonia being produced).

    d.   the temperature is increased for an endothermic reaction

Consider the reaction: $2NaHCO_3 + heat \rightleftharpoons Na_2CO_3 + H_2O + CO_2$

Although a change in temperature actually does change the *value* of the equilibrium constant, we can simplify reactions involving temperature changes by treating heat energy as if it were a chemical substance: for this endothermic reaction, heat is one of the reactants. As we saw in the example in part (a) of this question, increasing the concentration of one of the reactants for a system at equilibrium causes the reaction to proceed further to the right, forming additional product. Similarly for the endothermic reaction given above, increasing the temperature causes the reaction to proceed further in the direction of products than if no change had been made. It is as if there were too much "heat" to be compatible with the amount of substances present. The substances react to get rid of some of the energy.

    e.   the temperature is decreased for an exothermic process.

Consider the reaction: $PCl_3 + Cl_2 \rightleftharpoons PCl_5 + heat$

As discussed in part (d) above, although changing the temperature at which a reaction is performed does change the numerical value of K, we can simplify our discussion of this reaction by treating heat energy as if it were a chemical substance. Heat is a product of this reaction. If we are going to lower the temperature of this reaction system, the only way to accomplish this is to remove energy from the system. Lowering the temperature of the system is really working with this system in its attempt to release heat energy. So lowering the temperature should favor the production of more product than if no change were made.

16.
- a. HSO_3^-
- b. H_2SO_3
- c. HF
- d. CH_3COOH
- e. H_3O^+
- f. HNO_3
- g. H_2SO_4
- h. HS^-
- i. H_2S

224 *Cumulative Review: Chapters 15 and 16*

18. Bases (from Question 16)

    $SO_3^{2-} + H_2O \rightleftarrows HSO_3^- + OH^-$

    $F^- + H_2O \rightleftarrows HF + OH^-$

    $CH_3COO^- + H_2O \rightleftarrows CH_3COOH + OH^-$

    $H_2O + H_2O \rightleftarrows H_3O^+ + OH^-$

    $NO_3^- + H_2O \rightarrow$ (no reaction)

    $S^{2-} + H_2O \rightleftarrows HS^- + OH^-$

    $HS^- + H_2O \rightleftarrows H_2S + OH^-$

    Note: HSO_4^- and HSO_3^- are stronger acids than water and do not behave as bases in water

    Acids (from Question 17)

    $HC_2H_3O_2 + H_2O \rightleftarrows C_2H_3O_2^- + H_3O^+$

    $H_2S + H_2O \rightleftarrows HS^- + H_3O^+$

    $HS^- + H_2O \rightleftarrows S^{2-} + H_3O^+$

    $H_2CO_3 + H_2O \rightleftarrows HCO_3^- + H_3O^+$

    $HSO_4^- + H_2O \rightleftarrows SO_4^{2-} + H_3O^+$

    $H_2SO_4 + H_2O \rightarrow HSO_4^- + H_3O^+$

    $H_3PO_4 + H_2O \rightarrow H_2PO_4^- + H_3O^+$

    Note: HCO_3^- and NH_3 are stronger bases than water, and do not behave as acids in water; see also above.

20. a. HNO_3 is a strong acid, so

    $[H^+] = 0.00141\ M$

    $pH = -\log(0.00141) = 2.851$

    $pOH = 14 - 2.851 = 11.15$

    b. NaOH is a strong base, so

    $[OH^-] = 2.13 \times 10^{-3}\ M$

$$pOH = -\log(2.13 \times 10^{-3}) = 2.672$$

$$pH = 14 - 2.672 = 11.33$$

c. HCl is a strong acid, so

$$[H^+] = 0.00515 \; M$$

$$pH = -\log(0.00515) = 2.288$$

$$pOH = 14 - 2.288 = 11.71$$

d. $Ca(OH)_2$ is a strong, but not very soluble base. Each formula unit of $Ca(OH)_2$ produces two formula units of OH^- ion.

$$[OH^-] = 2 \times 5.65 \times 10^{-5} \; M = 1.13 \times 10^{-4} \; M$$

$$pOH = -\log(1.13 \times 10^{-4}) = 3.947$$

$$pH = 14 - 3.947 = 10.05$$

22. $2SO_2(g) + O_2(g) \rightleftarrows 2SO_3(g)$

$$K = \frac{[SO_3]^2}{[SO_2]^2[O_2]} = \frac{[0.42]^2}{[1.4 \times 10^{-3}]^2 [4.5 \times 10^{-4}]} = 2.0 \times 10^8$$

Chapter 17   Oxidation-Reduction Reactions/Electrochemistry

2. Reduction can be defined as the gaining of electrons by an atom, molecule, or ion. Reduction may also be defined as a decrease in oxidation state for an element, but naturally such a decrease takes place by the gaining of electrons (so the two definitions are equivalent).

   $S + 2e^- \rightarrow S$ is an example of a reduction process.

   $Na \rightarrow Na^+ + e^-$ is an example of an oxidation process.

4. Each of these reactions involves a *metallic* element in the form of the *free* element on one side of the equation; on the other side of the equation, the metallic element is *combined* in an ionic compound. If a metallic element goes from the free metal to the ionic form, the metal is oxidized (loses electrons).

   a. sodium is oxidized, nitrogen is reduced

   b. magnesium is oxidized, chlorine is reduced

   c. aluminum is oxidized, bromine is reduced

   d. iron is oxidized, oxygen is reduced

6. Most of these reactions involve a *metallic* element in the form of the *free* element on one side of the equation; on the other side of the equation, the metallic element is *combined* in an ionic compound. If a metallic element goes from the free metal to the ionic form, the metal is oxidized (loses electrons).

   a. magnesium is oxidized, bromine is reduced

   b. sodium is oxidized, sulfur is reduced

   c. bromide ion is oxidized, chlorine is reduced

   d. potassium is oxidized, nitrogen is reduced

8. The oxidation state of a *pure element* is *zero*, regardless of whether the element occurs naturally as single atoms or as a molecule.

10. Fluorine is always assigned a negative oxidation state (-1) because all other elements are less electronegative than fluorine. The other halogens are *usually* assigned an oxidation state of -1 in compounds. In interhalogen compounds such as ClF, fluorine is assigned oxidation state -1 (F is more electronegative than Cl), which means that chlorine must be assigned a +1 oxidation state in this instance.

12. Oxidation states represent a bookkeeping method to assign electrons in a molecule or ion. Since an ion has a net charge, the sum of the oxidation states of the atoms in the ion must equal the charge on the ion. For example, the hydroxide ion (OH^-) has an overall charge of -1 because hydrogen has an oxidation state of +1, whereas oxygen has an oxidation state of -2 in the hydroxide ion: (-2) + (+1) = -1

Chapter 17 Oxidation-Reduction Reactions/Electrochemistry   227

14. The rules for assigning oxidation states are given in Section 17.2 of the text. The rule which applies for each element in the following answers is given in parentheses after the element and its oxidation state.

   a.   N, +3 (Rule 6); Cl, -1 (Rule 5)

   b.   S, +6 (Rule 6); F, -1 (Rule 5)

   c.   P, +5 (Rule 6); Cl, -1 (Rule 5)

   d.   Si, -4 (Rule 6); H, +1 (Rule 4)

16. The rules for assigning oxidation states are given in Section 17.2 of the text. The rule which applies for each element in the following answers is given in parentheses after the element and its oxidation state.

   a.   H, +1 (Rule 4); Br, -1 (Rule 6 or Rule 5)

   b.   H, +1 (Rule 4); O, -2 (Rule 3); Br, +1 (Rule 6)

   c.   Br, 0 (Rule 1)

   d.   H, +1 (Rule 4); O, -2 (Rule 3); Br, +7 (Rule 6)

18. The rules for assigning oxidation states are given in Section 17.2 of the text. The rule which applies for each element in the following answers is given in parentheses after the element and its oxidation state.

   a.   H, +1 (Rule 4); O, -2 (Rule 3); N, +5 (Rule 6)

   b.   H, +1 (Rule 4); O, -2 (Rule 3); P, +5 (Rule 7)

   c.   H, +1 (Rule 4); O, -2 (Rule 3); S, +6 (Rule 7)

   d.   O, -1 (Rule 3)

20. The rules for assigning oxidation states are given in Section 17.2 of the text. The rule which applies for each element in the following answers is given in parentheses after the element and its oxidation state.

   a.   Cu, +2 (Rules 2,6); Cl, -1 (Rules 2,5)

   b.   Cr, +3 (Rules 2,6); Cl, -1 (Rules 2,5)

   c.   H, +1 (Rule 4); O, -2 (Rule 3); Cr, +6 (Rule 7)

   d.   Cr, +3 (Rules 2,6); O, -2 (Rules 2,3)

22. The rules for assigning oxidation states are given in Section 17.2 of the text. The rule which applies for each element in the following answers is given in parentheses after the element and its oxidation state.

a. H, +1 (Rule 4); C, −4 (Rule 6)

b. Na, +1 (Rule 2); O, −2 (Rule 3); C, +4 (Rule 6)

c. K, +1 (Rule 2); H, +1 (Rule 4); O, −2 (Rule 3); C, +4 (Rule 6)

d. O, −2 (Rule 3); C, +2 (Rule 6)

24. Electrons are negative; when an atom gains electrons, it gains one negative charge for each electron gained. For example, in the reduction reaction $Cl + e^- \rightarrow Cl^-$, the oxidation state of chlorine decreases from 0 to −1 as the electron is gained.

26. An oxidizing agent *causes* another species to be oxidized (to lose electrons). In order to make another species lose electrons, the oxidizing agent must be capable of gaining the electrons; an oxidizing agent is itself reduced. On the contrary, a reducing agent is itself oxidized.

28. An oxidizing agent oxidizes another species by gaining the electrons lost by the other species; therefore, an oxidizing agent itself decreases in oxidation state. A reducing agent increases its oxidation state when acting on another atom or molecule.

30. a. $Zn(s) + 2HNO_3(g) \rightarrow Zn(NO_3)_2(aq) + H_2(g)$

    Zn 0   H +1   Zn +2   H 0

              N +5   N +5

              O −2   O −2

zinc is oxidized; hydrogen is reduced

b. $H_2(g) + CuSO_4(aq) \rightarrow Cu(s) + H_2SO_4(aq)$

    H 0   Cu +2   Cu 0   H +1

           S +6   S +6

           O −2   O −2

hydrogen is oxidized; copper is reduced

c. $N_2(g) + 3Br_2(l) \rightarrow 2NBr_3(g)$

    N 0   Br 0   N −3

                  Br +1

N is more electronegative than Br

nitrogen is reduced; bromine is oxidized

d. $2KBr(aq) + Cl_2(g) \rightarrow 2KCl(aq) + Br_2(l)$

    K +1   Cl 0   K +1   Br 0

    Br −1   Cl −1

bromine is oxidized; chlorine is reduced

32. a. $Cu(s) + 2AgNO_3(aq) \rightarrow 2Ag(s) + Cu(NO_3)_2(aq)$

    Cu 0   Ag +1   Ag 0   Cu +2

             N +5   N +5

             O −2   O −2

copper is oxidized; silver is reduced

b. $N_2(g) + 3F_2(g) \rightarrow 2NF_3(g)$

    N 0   F 0   F −1

                      N +3

nitrogen is oxidized; fluorine is reduced

c. $2Fe_2O_3(s) + 3S(s) \rightarrow 4Fe(s) + 3SO_2$

    Fe +3   S 0   Fe 0   S +4

    O −2   O −2

sulfur is oxidized; iron is reduced

d. $2H_2O_2(l) \rightarrow 2H_2O(l) + O_2(g)$

    H +1   H +1   O 0

    O −1   O −2

oxygen is both oxidized and reduced

34. Iron is reduced [+3 in $Fe_2O_3(s)$, 0 in $Fe(l)$]; carbon is oxidized [+2 in $CO(g)$, +4 in $CO_2(g)$]. $Fe_2O_3(s)$ is the oxidizing agent; $CO(g)$ is the reducing agent.

36. Chlorine is reduced [0 in $Cl_2(g)$, −1 in $NaCl(s)$]; bromine is oxidized [−1 in $NaBr(aq)$, 0 in $Br_2(l)$]. $Cl_2(g)$ is the oxidizing agent; $NaBr(aq)$ is the reducing agent.

38. Oxidation-reduction reactions are often more complicated than "regular" reactions; frequently the coefficients necessary to balance the number of electrons transferred come out to be large numbers. We also have to make certain that we account for the electrons being transferred.

40. Under ordinary conditions it is impossible to have "free" electrons that are not part of some atom, ion, or molecule. For this reason, the total

number of electrons lost by the species being oxidized must equal the total number of electrons gained by the species being reduced.

42. a. $N_2(g) \rightarrow N^{3-}(s)$

   Balance nitrogen: $N_2(g) \rightarrow \mathbf{2N^{3-}(s)}$

   Balance charge: $\mathbf{6e^-} + N_2(g) \rightarrow 2N^{3-}(s)$

   Balanced half reaction: $6e^- + N_2(g) \rightarrow 2N^{3-}(s)$

   b. $O_2^{2-}(aq) \rightarrow O_2(g)$

   Balance charge: $O_2^{2-}(aq) \rightarrow O_2(g) + \mathbf{2e^-}$

   Balanced half reaction: $O_2^{2-}(aq) \rightarrow O_2(g) + 2e^-$

   c. $Zn(s) \rightarrow Zn^{2+}(aq)$

   Balance charge: $Zn(s) \rightarrow Zn^{2+}(aq) + \mathbf{2e^-}$

   Balanced half reaction: $Zn(s) \rightarrow Zn^{2+}(aq) + 2e^-$

   d. $F_2(g) \rightarrow F^-(aq)$

   Balance fluorine: $F_2(g) \rightarrow \mathbf{2F^-(aq)}$

   Balance charge: $\mathbf{2e^-} + F_2(g) \rightarrow 2F^-(aq)$

   Balanced half reaction: $2e^- + F_2(g) \rightarrow 2F^-(aq)$

44. a. $O_2(g) \rightarrow H_2O(l)$

   Balance oxygen: $O_2(g) \rightarrow \mathbf{2H_2O(l)}$

   Balance hydrogen: $\mathbf{4H^+}(aq) + O_2(g) \rightarrow 2H_2O(l)$

   Balance charge: $\mathbf{4e^-} + 4H^+(aq) + O_2(g) \rightarrow 2H_2O(l)$

   Balanced half reaction: $4e^- + 4H^+(aq) + O_2(g) \rightarrow 2H_2O(l)$

   b. $IO_3^-(aq) \rightarrow I_2(s)$

   Balance iodine: $\mathbf{2}IO_3^-(aq) \rightarrow I_2(s)$

   Balance oxygen: $2IO_3^-(aq) \rightarrow I_2(s) + \mathbf{6H_2O(l)}$

   Balance hydrogen: $\mathbf{12H^+} + 2IO_3^-(aq) \rightarrow I_2(s) + 6H_2O(l)$

   Balance charge: $12H^+(aq) + 2IO_3^-(aq) + \mathbf{10e^-} \rightarrow I_2(s) + 6H_2O(l)$

   Balanced half reaction: $12H^+(aq) + 2IO_3^-(aq) + 10e^- \rightarrow I_2(s) + 6H_2O(l)$

   c. $VO^{2+}(aq) \rightarrow V^{3+}(aq)$

   Balance oxygen: $VO^{2+}(aq) \rightarrow V^{3+}(aq) + \mathbf{H_2O(l)}$

   Balance hydrogen: $\mathbf{2H^+}(aq) + VO^{2+}(aq) \rightarrow V^{3+}(aq) + H_2O(l)$

Balance charge: $e^- + 2H^+(aq) + VO^{2+}(aq) \rightarrow V^{3+}(aq) + H_2O(l)$

Balanced half reaction: $e^- + 2H^+(aq) + VO^{2+}(aq) \rightarrow V^{3+}(aq) + H_2O(l)$

d. $BiO^+(aq) \rightarrow Bi(s)$

Balance oxygen: $BiO^+(aq) \rightarrow Bi(s) + \mathbf{H_2O}(l)$

Balance hydrogen: $\mathbf{2H^+}(aq) + BiO^+(aq) \rightarrow Bi(s) + H_2O(l)$

Balance charge: $\mathbf{3e^-} + 2H^+(aq) + BiO^+(aq) \rightarrow Bi(s) + H_2O(l)$

Balanced half reaction: $3e^- + 2H^+(aq) + BiO^+(aq) \rightarrow Bi(s) + H_2O(l)$

46. For simplicity, the physical states of the substances have been omitted until the final balanced equation is given.

    a. $MnO_4^-(aq) + Zn(s) \rightarrow Mn^{2+}(aq) + Zn^{2+}(aq)$

    $MnO_4^- \rightarrow Mn^{2+}$

    Balance oxygen: $MnO_4^- \rightarrow Mn^{2+} + \mathbf{4H_2O}$

    Balance hydrogen: $\mathbf{8H^+} + MnO_4^- \rightarrow Mn^{2+} + 4H_2O$

    Balance charge: $\mathbf{5e^-} + 8H^+ + MnO_4^- \rightarrow Mn^{2+} + 4H_2O$

    $Zn \rightarrow Zn^{2+}$

    Balance charge: $Zn \rightarrow Zn^{2+} + \mathbf{2e^-}$

    Combine half reactions: $2 \times (5e^- + 8H^+ + MnO_4^- \rightarrow Mn^{2+} + 4H_2O)$
    $5 \times (Zn \rightarrow Zn^{2+} + 2e^-)$

    $16H^+(aq) + 2MnO_4^-(aq) + 5Zn(s) \rightarrow 2Mn^{2+}(aq) + 8H_2O(l) + 5Zn^{2+}(aq)$

    b. $Sn^{4+}(aq) + H_2(g) \rightarrow Sn^{2+}(aq) + H^+(aq)$

    $Sn^{4+} \rightarrow Sn^{2+}$

    Balance charge: $Sn^{4+} + \mathbf{2e^-} \rightarrow Sn^{2+}$

    $H_2 \rightarrow H^+$

    Balance hydrogen: $H_2 \rightarrow \mathbf{2H^+}$

    Balance charge: $H_2 \rightarrow 2H^+ + \mathbf{2e^-}$

    $Sn^{4+}(aq) + H_2(g) \rightarrow Sn^{2+}(aq) + 2H^+(aq)$

c. $Zn(s) + NO_3^-(aq) \rightarrow Zn^{2+}(aq) + NO_2(g)$

$Zn \rightarrow Zn^{2+}$

Balance charge: $Zn \rightarrow Zn^{2+} + \mathbf{2e^-}$

$NO_3^- \rightarrow NO_2$

Balance oxygen: $NO_3^- \rightarrow NO_2 + \mathbf{H_2O}$

Balance hydrogen: $NO_3^- + \mathbf{2H^+} \rightarrow NO_2 + H_2O$

Balance charge: $NO_3^- + 2H^+ + \mathbf{e^-} \rightarrow NO_2 + H_2O$

Balanced half-reaction: $NO_3^- + 2H^+ + e^- \rightarrow NO_2 + H_2O$

Combine half reactions: $Zn \rightarrow Zn^{2+} + 2e^-$

$2 \times (NO_3^- + 2H^+ + e^- \rightarrow NO_2 + H_2O)$

$Zn(s) + 2NO_3^-(aq) + 4H^+(aq) \rightarrow Zn^{2+}(aq) + 2NO_2(g) + 2H_2O(l)$

d. $H_2S(g) + Br_2(l) \rightarrow S(s) + Br^-(aq)$

$H_2S \rightarrow S$

Balance hydrogen: $H_2S \rightarrow S + \mathbf{2H^+}$

Balance charge: $H_2S \rightarrow S + 2H^+ + \mathbf{2e^-}$

Balanced half-reaction: $H_2S \rightarrow S + 2H^+ + 2e^-$

$Br_2 \rightarrow Br^-$

Balance bromine: $Br_2 \rightarrow 2Br^-$

Balance charge: $\mathbf{2e^-} + Br_2 \rightarrow 2Br^-$

Balanced half-reaction: $2e^- + Br_2 \rightarrow 2Br^-$

$H_2S(g) + Br_2(l) \rightarrow S(s) + 2H^+(aq) + 2Br^-(aq)$

48. The half reaction for the reduction of Ce^{4+} is the *same* in each of these processes:

$Ce^{4+} + e^- \rightarrow Ce^{3+}$

a. $H_3AsO_3 \rightarrow H_3AsO_4$

Balance oxygen: $\mathbf{H_2O} + H_3AsO_3 \rightarrow H_3AsO_4$

Balance hydrogen: $H_2O + H_3AsO_3 \rightarrow H_3AsO_4 + \mathbf{2H^+}$

Balance charge: $H_2O + H_3AsO_3 \rightarrow H_3AsO_4 + 2H^+ + \mathbf{2e^-}$

Balanced half reaction: $H_2O + H_3AsO_3 \rightarrow H_3AsO_4 + 2H^+ + 2e^-$

Combine half reactions: $H_2O + H_3AsO_3 \rightarrow H_3AsO_4 + 2H^+ + 2e^-$
$2 \times (Ce^{4+} + e^- \rightarrow Ce^{3+})$

$H_2O(l) + H_3AsO_3(aq) + 2Ce^{4+}(aq) \rightarrow H_3AsO_4(aq) + 2H^+(aq) + 2Ce^{3+}(aq)$

b. $Fe^{2+} \rightarrow Fe^{3+}$

Balance charge: $Fe^{2+} \rightarrow Fe^{3+} + $ **e^-**

Balanced half reaction: $Fe^{2+} \rightarrow Fe^{3+} + e^-$

Combine half reactions: $Fe^{2+} \rightarrow Fe^{3+} + e^-$
$Ce^{4+} + e^- \rightarrow Ce^{3+}$

$Ce^{4+}(aq) + Fe^{2+}(aq) \rightarrow Ce^{3+}(aq) + Fe^{3+}(aq)$

c. $I^- \rightarrow I_2$

Balance iodine: **$2I^-$** $\rightarrow I_2$

Balance charge: $2I^- \rightarrow I_2 + $ **$2e^-$**

Balanced half reaction: $2I^- \rightarrow I_2 + 2e^-$

Combine half reactions: $2I^- \rightarrow I_2 + 2e^-$
$2 \times (Ce^{4+} + e^- \rightarrow Ce^{3+})$

$2Ce^{4+}(aq) + 2I^-(aq) \rightarrow 2Ce^{3+}(aq) + I_2(s)$

50. A salt bridge typically consists of a U-shaped tube filled with an inert electrolyte (one involving ions that are not part of the oxidation-reduction reaction). A salt bridge is used to complete the electrical circuit in a cell. Any method which allows transfer of charge without allowing bulk mixing of the solutions may be used (another common method is to set up one half-cell in a porous cup, which is then placed in the beaker containing the second half-cell).

52. In a galvanic cell, the anode is the electrode where oxidation occurs; the cathode is the electrode where reduction occurs.

54.

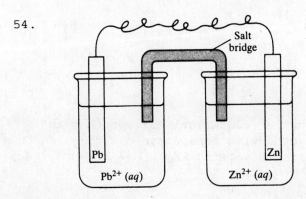

$Pb^{2+}(aq)$ ion is reduced
$Zn(s)$ is oxidized.
anode reaction: $Zn(s) \rightarrow Zn^{2+}(aq) + 2e^-$.
cathode rection $Pb^{2+}(aq) + 2e^- \rightarrow Pb(s)$

234   Chapter 17   Oxidation-Reduction Reactions/Electrochemistry

56. Both normal and alkaline cells contain zinc as one electrode; zinc corrodes more slowly under alkaline conditions than in the highly acidic environment of a normal dry cell.

   anode: $Zn(s) + 2OH^-(aq) \rightarrow ZnO(s) + H_2O(l) + 2e^-$

   cathode: $2MnO_2(s) + H_2O(l) + 2e^- \rightarrow Mn_2O_3(s) + 2OH^-(aq)$

58. Aluminum is a very reactive metal when freshly isolated in the pure state. However, on standing for even a relatively short period of time, aluminum metal forms a thin coating of Al_2O_3 on its surface from reaction with atmospheric oxygen. This coating of Al_2O_3 is much less reactive than the metal and serves to protect the surface of the metal from further attack.

60. In cathodic protection of steel tanks and pipes, a more reactive metal than iron is connected to the item to be protected. The active metal is then preferentially oxidized rather than the iron of the tank or pipe.

62. The main recharging reaction for the lead storage battery is

    $2PbSO_4(s) + 2H_2O(l) \rightarrow Pb(s) + PbO_2(s) + 2H_2SO_4(aq)$

    A major side reaction is the electrolysis of water

    $2H_2O(l) \rightarrow 2H_2(g) + O_2(g)$

    resulting in the production of an explosive mixture of hydrogen and oxygen, which accounts for many accidents during the recharging of such batteries.

64. Electrolysis is applied in electroplating by making the item to be plated the cathode in a cell containing a solution of ions of the desired plating metal.

66. loss; oxidation state

68. electronegative

70. An *oxidizing agent* is an atom, molecule, or ion which causes the oxidation of another species. During this process, the oxidizing agent itself is reduced.

72. lose

74. separate from

76. oxidation

78. An electrolysis reaction results when an electrical current from an outside source is used to cause an otherwise nonspontaneous reaction to occur. An example is the electrolysis of water: $2H_2O(l) \rightarrow 2H_2(g) + O_2(g)$;

Chapter 17   Oxidation-Reduction Reactions/Electrochemistry   235

this reaction only takes place if an electrical current of sufficient voltage is passed through the water.

80. hydrogen; oxygen

82. oxidation

84.
a. $4Fe(s) + 3O_2(g) \rightarrow 2Fe_2O_3(s)$
iron is oxidized, oxygen is reduced

b. $2Al(s) + 3Cl_2(g) \rightarrow 2AlCl_3(s)$
aluminum is oxidized, chlorine is reduced

c. $6Mg(s) + P_4(s) \rightarrow 2Mg_3P_2(s)$
magnesium is oxidized, phosphorus is reduced

86.
a. aluminum is oxidized; hydrogen is reduced
b. hydrogen is reduced; iodine is oxidized
c. copper is oxidized; hydrogen is reduced

88.
a. $C_3H_8(g) + 5O_2(g) \rightarrow 3CO_2(g) + 4H_2O(g)$
b. $CO(g) + 2H_2(g) \rightarrow CH_3OH(l)$
c. $SnO_2(s) + 2C(s) \rightarrow Sn(s) + 2CO(g)$
d. $C_2H_5OH(l) + 3O_2(g) \rightarrow 2CO_2(g) + 3H_2O(g)$

90. Each of these reactions involves a *metallic* element in the form of the *free* element on one side of the equation; on the other side of the equation, the metallic element is *combined* in an ionic compound. If a metallic element goes from the free metal to the ionic form, the metal is oxidized (loses electrons).

a. sodium is oxidized, oxygen is reduced
b. iron is oxidized, hydrogen is reduced
c. oxygen (O^{2-}) is oxidized, aluminum (Al^{3+}) is reduced (this reaction is the reverse of the type discussed above)
d. magnesium is oxidized, nitrogen is reduced

92. The rules for assigning oxidation states are given in Section 17.2 of the text. The rule which applies for each element in the following answers is given in parentheses after the element and its oxidation state.

a. H +1 (Rule 4); N −3 (Rule 6)
b. C +2 (Rule 6); O −2 (Rule 3)

236   Chapter 17   Oxidation-Reduction Reactions/Electrochemistry

    c.   C +4 (Rule 6); O -2 (Rule 3)

    d.   N +3 (Rule 6); F -1 (Rule 5)

94.   The rules for assigning oxidation states are given in Section 17.2 of the text. The rule which applies for each element in the following answers is given in parentheses after the element and its oxidation state.

    a.   Mn +4 (Rule 6); O -2 (Rule 3)

    b.   Ba +2 (Rule 2); Cr +6 (Rule 6); O -2 (Rule 3)

    c.   H +1 (Rule 4); S +4 (Rule 6); O -2 (Rule 3)

    d.   Ca +2 (Rule 2); P +5 (Rule 6); O -2 (Rule 3)

96.   The rules for assigning oxidation states are given in Section 17.2 of the text. The rule which applies for each element in the following answers is given in parentheses after the element and its oxidation state.

    a.   Bi +3 (Rule 7); O -2 (Rule 3)

    b.   P +5 (Rule 7); O -2 (Rule 3)

    c.   N +3 (Rule 7); O -2 (Rule 3)

    d.   Hg +1 (Rule 7)

98.   a. $2B_2O_3(s) + 6Cl_2(g) \rightarrow 4BCl_3(l) + 3O_2(g)$

           B +3   Cl 0   B +3   O 0

           O -2   Cl -1

        oxygen is oxidized; chlorine is reduced

    b. $GeH_4(g) + O_2(g) \rightarrow Ge(s) + 2H_2O(g)$

           Ge -4   O 0   Ge 0   H +1

           H +1   O -2

        germanium is oxidized; oxygen is reduced

    c. $C_2H_4(g) + Cl_2(g) \rightarrow C_2H_4Cl_2(l)$

           C -2   Cl 0   C -1

           H +1   H +1; Cl -1

        carbon is oxidized; chlorine is reduced

    d. $O_2(g) + 2F_2(g) \rightarrow 2OF_2(g)$

           O 0   F 0   O +2

                                  F -1

        oxygen is oxidized; fluorine is reduced

100. a. $SiO_2(s) \rightarrow Si(s)$

Balance oxygen: $SiO_2(s) \rightarrow Si(s) + \mathbf{2H_2O}(l)$

Balance hydrogen: $SiO_2(s) + \mathbf{4H^+}(aq) \rightarrow Si(s) + 2H_2O(l)$

Balance charge: $SiO_2(s) + 4H^+(aq) + \mathbf{4e^-} \rightarrow Si(s) + 2H_2O(l)$

Balanced half reaction: $SiO_2(s) + 4H^+(aq) + 4e^- \rightarrow Si(s) + 2H_2O(l)$

b. $S(s) \rightarrow H_2S(g)$

Balance hydrogen: $S(s) + \mathbf{2H^+}(aq) \rightarrow H_2S(g)$

Balance charge: $S(s) + 2H^+(aq) + \mathbf{2e^-} \rightarrow H_2S(g)$

Balanced half reaction: $S(s) + 2H^+(aq) + 2e^- \rightarrow H_2S(g)$

c. $NO_3^-(aq) \rightarrow HNO_2(aq)$

Balance oxygen: $NO_3^-(aq) \rightarrow HNO_2(aq) + \mathbf{H_2O}(l)$

Balance hydrogen: $NO_3^-(aq) + \mathbf{3H^+}(aq) \rightarrow HNO_2(aq) + H_2O(l)$

Balance charge: $NO_3^-(aq) + 3H^+(aq) + \mathbf{2e^-} \rightarrow HNO_2(aq) + H_2O(l)$

Balanced half reaction: $NO_3^-(aq) + 3H^+(aq) + 2e^- \rightarrow HNO_2(aq) + H_2O(l)$

d. $NO_3^-(aq) \rightarrow NO(g)$

Balance oxygen: $NO_3^-(aq) \rightarrow NO(g) + \mathbf{2H_2O}(l)$

Balance hydrogen: $NO_3^-(aq) + \mathbf{4H^+}(aq) \rightarrow NO(g) + 2H_2O(l)$

Balance charge: $NO_3^-(aq) + 4H^+(aq) + \mathbf{3e^-} \rightarrow NO(g) + 2H_2O(l)$

Balanced half reaction: $NO_3^-(aq) + 4H^+(aq) + 3e^- \rightarrow NO(g) + 2H_2O(l)$

102. For simplicity, the physical states of the substances have been omitted until the final balanced equation is given.

For the reduction of the permanganate ion, MnO_4^-, in acid solution, the half reaction is always the *same*:

$MnO_4^- \rightarrow Mn^{2+}$

Balance oxygen: $MnO_4^- \rightarrow Mn^{2+} + \mathbf{4H_2O}$

Balance hydrogen: $\mathbf{8H^+} + MnO_4^- \rightarrow Mn^{2+} + 4H_2O$

Balance charge: $8H^+ + MnO_4^- + \mathbf{5e^-} \rightarrow Mn^{2+} + 4H_2O$

a. $C_2O_4^{2-} \rightarrow CO_2$

Balance carbon: $C_2O_4^{2-} \rightarrow \mathbf{2CO_2}$

Balance charge: $C_2O_4^{2-} \rightarrow 2CO_2 + \mathbf{2e^-}$

Combine half reactions: $5 \times (C_2O_4^{2-} \rightarrow \mathbf{2}CO_2 + 2e^-)$

$ 2 \times (8H^+ + MnO_4^- + 5e^- \rightarrow Mn^{2+} + 4H_2O)$

$16H^+(aq) + 2MnO_4^-(aq) + 5C_2O_4^{2-}(aq) \rightarrow 2Mn^{2+}(aq) + 8H_2O(l) + 10CO_2(g)$

b. $Fe^{2+} \rightarrow Fe^{3+}$

Balance charge: $Fe^{2+} \rightarrow Fe^{3+} + \mathbf{e^-}$

Combine half reactions: $5 \times (Fe^{2+} \rightarrow Fe^{3+} + e^-)$

$ 8H^+ + MnO_4^- + 5e^- \rightarrow Mn^{2+} + 4H_2O$

$8H^+(aq) + MnO_4^-(aq) + 5Fe^{2+}(aq) \rightarrow Mn^{2+}(aq) + 4H_2O(l) + 5Fe^{3+}(aq)$

c. $Cl^- \rightarrow Cl_2$

Balance chlorine: $\mathbf{2}Cl^- \rightarrow Cl_2$

Balance charge: $2Cl^- \rightarrow Cl_2 + \mathbf{2e^-}$

Combine half reactions: $5 \times (2Cl^- \rightarrow Cl_2 + 2e^-)$

$ 2 \times (8H^+ + MnO_4^- + 5e^- \rightarrow Mn^{2+} + 4H_2O)$

$16H^+(aq) + 2MnO_4^-(aq) + 10Cl^-(aq) \rightarrow 2Mn^{2+}(aq) + 8H_2O(l) + 5Cl_2(g)$

Chapter 18   Radioactivity and Nuclear Energy

2. The radius of a typical atomic nucleus is on the order of 10^{-13} cm, which is about one hundred thousand times smaller than the radius of an atom overall.

4. The atomic number (Z) of a nucleus represents the number of protons present in the nucleus. The mass number (A) of a nucleus represents the total number of protons and neutrons in the nucleus. For example, for the nuclide $^{13}_{6}C$, with six protons and seven neutrons, we have $Z = 6$ and $A = 13$.

6. The atomic number (Z) is written in such formulas as a left subscript, while the mass number (A) is written as a left superscript. That is, the general symbol for a nuclide is $^{A}_{Z}X$. As an example, consider the isotope of oxygen with 8 protons and 8 neutrons: its symbol would be $^{16}_{8}O$.

8. Alpha particle: charge 2+, mass number 4, symbol $^{4}_{2}He$

10. When a nucleus produces a beta (β) particle, the atomic number of the parent nucleus is *increased* by *one* unit. The beta particle has a mass number of zero, but an "atomic number" of –1.

12. Gamma rays are high energy photons of electromagnetic radiation. Gamma rays are not considered to be particles. When a nucleus produces only gamma radiation, the atomic number and mass number of the nucleus do not change. Gamma rays represent the energy changes associated with transitions and rearrangement of the particles within the nucleus.

14. Electron capture occurs when one of the inner orbital electrons is pulled into and becomes part of the nucleus.

16. The average atomic mass listed on the Periodic Table is 20.18, which suggests that the isotope of mass number 20 predominates in naturally occurring neon. The average atomic mass given in the periodic table is a *weighted* average which includes not only the masses of the isotopes, but also their relative abundances.

18. The approximate atomic molar mass could be calculated as follows: $0.79(24) + 0.10(25) + 0.11(26) = 24.3$. This is *only* and approximation since the mass numbers, rather than the actual isotopic masses, were used. The fact that the approximate mass calculated is slightly above 24 shows that the isotope of mass number 24 predominates.

20. a. $^{0}_{-1}e$ or $^{0}_{-1}\beta$

    b. $^{0}_{+1}e$ or $^{0}_{+1}\beta$

    c. $^{0}_{0}\gamma$

22. a. $^{192}_{83}Bi$

b. $^{204}_{82}Pb$

c. $^{206}_{84}Po$

24. a. $^{0}_{-1}e$

b. $^{0}_{-1}e$

c. $^{210}_{83}Bi$

26. a. $^{136}_{53}I \rightarrow {}^{136}_{54}Xe + {}^{0}_{-1}e$

b. $^{133}_{51}Sb \rightarrow {}^{133}_{52}Te + {}^{0}_{-1}e$

c. $^{117}_{49}In \rightarrow {}^{117}_{50}Sn + {}^{0}_{-1}e$

d. $^{47}_{20}Ca \rightarrow {}^{47}_{21}Sc + {}^{0}_{-1}e$

28. a. $^{226}_{88}Ra \rightarrow {}^{222}_{86}Rn + {}^{4}_{2}He$

b. $^{222}_{86}Rn \rightarrow {}^{218}_{84}Po + {}^{4}_{2}He$

c. $^{239}_{94}Pu \rightarrow {}^{235}_{92}U + {}^{4}_{2}He$

d. $^{8}_{4}Be \rightarrow {}^{4}_{2}He + {}^{4}_{2}He$

30. There is often considerable repulsion between the target nucleus and the particles being used for bombardment (especially if the bombarding particle is positively charged like the target nucleus). Using accelerators to greatly speed up the bombarding particles can overcome this repulsion.

32. $^{27}_{13}\text{Al} + ^{4}_{2}\text{He} \rightarrow ^{30}_{15}\text{P} + ^{1}_{0}\text{n}$

34. The half-life of a nucleus is the time required for one-half of the original sample of nuclei to decay. A given isotope of an element always has the same half-life, although different isotopes of the same element may have greatly different half-lives. Nuclei of different elements typically have different half-lives.

36. $^{226}_{88}\text{Ra}$ is the most stable (longest half-life)

    $^{224}_{88}\text{Ra}$ is the "hottest" (shortest half-life)

38. highest activity   lowest activity

    $^{87}\text{Sr} > {}^{99}\text{Tc} > {}^{24}\text{Na} > {}^{99}\text{Mo} > {}^{133}\text{Xe} > {}^{131}\text{I} > {}^{32}\text{P} > {}^{51}\text{Cr} > {}^{59}\text{Fe}$

40. For ^{223}Ra, the half-life is 12 days. After two half-lives (24 days), 250 mg remains; after three half-lives (36 days), 125 mg remains.

    For ^{224}Ra, the half-life is 3.6 days. One month would be approximately 8 half-life periods (29 days), and approximately 4 mg remains.

    For ^{225}Ra, the half-life is 15 days. One month would be two half-life periods, and 250 mg remains.

42. For an administered dose of 100 μg, 0.39 μg remains after 2 days. The fraction remaining is 0.39/100 = 0.0039; on a percentage basis, less than 0.4 % of the original radioisotope remains.

44. Carbon-14 is produced in the upper atmosphere by the bombardment of ordinary nitrogen with neutrons from space:

    $^{14}_{7}\text{N} + ^{1}_{0}\text{n} \rightarrow ^{14}_{6}\text{C} + ^{1}_{1}\text{H}$

46. We assume that the concentration of C-14 in the atmosphere is effectively constant. A living organism is constantly replenishing C-14 either through the processes of metabolism (sugars ingested in foods contain C-14), or photosynthesis (carbon dioxide contains C-14). When a plant dies, it no longer replenishes itself with C-14 from the atmosphere, and as the C-14 undergoes radioactive decay, its amount decreases with time.

48. These isotopes and their uses are listed in Table 18.4 Some important examples include the use of I-131 (and other iodine isotopes) in the

diagnosis and treatment of thyroid disease (iodine is used in the body primarily in the thyroid gland); Fe-59 in the study of the function of red blood cells (iron is a constituent of hemoglobin which is found in the red blood cells); Sr-87 in the study of bones (Sr is a Group 2 element, and is able to take the place of Ca in bone structures).

50. Combining two light nuclei to form a heavier, more stable nucleus is called nuclear *fusion*. Splitting a heavy nucleus into nuclei with smaller mass numbers is called nuclear *fission*.

52. $^{1}_{0}n + ^{235}_{92}U \rightarrow ^{142}_{56}Ba + ^{91}_{36}Kr + 3\, ^{1}_{0}n$ is one possibility.

54. A critical mass of a fissionable material is the amount needed to provide a high enough internal neutron flux to sustain the chain reaction (enough neutrons are produced to cause the continuous fission of further material). A sample with less than a critical mass is still radioactive, but cannot sustain a chain reaction.

56. An actual nuclear explosion, of the type produced by a nuclear weapon, cannot occur in a nuclear reactor because the concentration of the fissionable materials is not sufficient to form a supercritical mass. However, since many reactors are cooled by water, which can decompose into hydrogen and oxygen gases, a *chemical* explosion is possible which could scatter the radioactive material used in the reactor.

58. $^{238}_{92}U + ^{1}_{0}n \rightarrow ^{239}_{92}U$

$^{239}_{92}U \rightarrow ^{239}_{93}Np + ^{0}_{-1}e$

$^{239}_{93}Np \rightarrow ^{239}_{94}Pu + ^{0}_{-1}e$

60. In one type of fusion reactor, two $^{2}_{1}H$ atoms are fused to produce $^{4}_{2}He$. Because the hydrogen nuclei are positively charged, extremely high energies (temperatures of 40 million K) are needed to overcome the repulsion between the nuclei as they are shot into each other.

62. In the theory of stellar nucleosynthesis, it is considered that the nucleus began as a cloud of neutrons which exploded (the Big Bang). After this initial explosion, neutrons were thought to have decomposed into protons and electrons

$^{1}_{0}n \rightarrow ^{1}_{1}H + ^{0}_{-1}e$

The products of this decomposition were then thought to have combined to form large clouds of hydrogen atoms. As the hydrogen clouds became larger, gravitational forces caused these clouds to contract and heat up. Eventually the clouds of hydrogen were so dense and so hot that fusion of hydrogen nuclei into helium nuclei took place, with a great release of energy. When the tendency for the hydrogen clouds to expand from the heat of fusion was counter-balanced by the gravitational forces of the cloud, a small star had formed. In addition to the fusion of hydrogen nuclei into helium mentioned already, as the star's hydrogen supply is exhausted, the helium present in the star also begins to undergo fusion into nuclei of other elements.

64. Somatic damage is damage directly to the organism itself, causing nearly immediate sickness or death to the organism. Genetic damage is damage to the genetic machinery of the organism, which will be manifested in future generations of offspring.

66. Gamma rays penetrate long distances, but seldom cause ionization of biological molecules. Alpha particles, because they are much heavier although less penetrating, are very effective at ionizing biological molecules and leave a dense trail of damage in the organism. Isotopes which release alpha particles can be ingested or breathed into the body where the damage from the alpha particles will be more acute.

68. The exposure limits given in Table 18.5 as causing no detectable clinical effect are 0-25 rem. The total yearly exposures from natural and human-induced radioactive sources are estimated in Table 18.6 as less than 200 *milli*rem (0.2 rem), which is well within the acceptable limits.

70. radioactive

72. mass

74. neutron; proton

76. radioactive decay

78. mass number

80. transuranium

82. half-life

84. radiotracers

86. chain

88. breeder

90. 4.5×10^9 dollars ($4.5 billion)

244   Chapter 18   Radioactivity and Nuclear Energy

92. 3.5×10^{-11} J/atom; 8.9×10^{10} J/g

94. Despite the fact that nuclear waste has been generated for over 40 years, no permanent disposal plan has been implemented as yet. One proposal to dispose of such waste calls for the waste to be sealed in blocks of glass, which in turn are sealed in corrosion-proof metal drums, which would then be buried in deep, stable rock formations away from earthquake and other geologically active zones. In these deep storage areas, it is hoped that the waste could decay safely undisturbed until the radioactivity drops to "safe" levels.

96. $^{27}_{13}$Al: 13 protons, 14 neutrons

    $^{28}_{13}$Al: 13 protons, 15 neutrons

    $^{29}_{13}$Al: 13 protons, 16 neutrons

98. a. $^{0}_{-1}e$

    b. $^{74}_{34}$Se

    c. $^{240}_{92}$U

100. For a decay of 10 μg to 1/1000 of this amount, we want to know when the amount of remaining ^{131}I is on the order of 0.01 μg.

time, days	0	8	16	24	32	40
mass, μg	10	5	2.5	1.25	0.625	0.313
time, days	48	56	64	72	80	
mass, μg	0.156	0.078	0.039	0.020	0.01	

Approximately 80 days are required.

102. Breeder reactors are set up to convert non-fissionable ^{238}U into fissionable ^{239}Pu. The material used for fission in a breeder reactor is a combination of U-235 (which undergoes fission in a chain reaction) and the more common U-238 isotope. Excess neutrons from the U-235 fission are absorbed by the U-238 converting it to the fissionable plutonium isotope Pu-239. Although Pu-239 is fissionable, its chemical and physical properties make it very difficult and expensive to handle and process.

Chapter 19   Organic Chemistry

2. A given carbon atom can be attached to a maximum of four other atoms. Carbon atoms have four valence electrons. By making four bonds, carbon atoms exactly complete their valence octet.

4. A triple bond represents the sharing of *six* electrons (three *pairs* of electrons). The simplest example of an organic molecule containing a triple bond is acetylene, H:C:::C:H (H-C≡C-H).

6. Each carbon atom in ethane is bonded to *four* other atoms. According to VSEPR theory, each carbon atom has its electron pairs arranged *tetrahedrally*.

8. An unsaturated hydrocarbon is one that contains an area of multiple bonding between some of the carbon atoms. Such a hydrocarbon is called "unsaturated" because each carbon atom of the multiple bond could bond to additional atoms (rather than to each other). Such unsaturated hydrocarbons undergo addition reactions in which atoms or groups from an outside reagent become bonded to the carbon atoms of the multiple bond, producing a saturated compound. The structural features that characterize unsaturated hydrocarbons are the carbon-carbon double bond or triple bond.

10. Each successive member of this family differs from the previous member by a -CH$_2$- unit (which is sometimes called a "methylene" unit). Such a family of compounds is sometimes referred to as a *homologous series*.

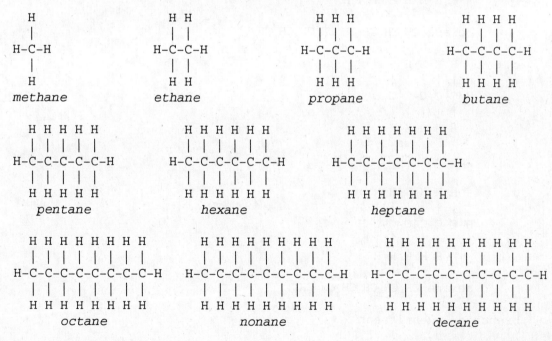

12. a.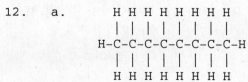

246   Chapter 19   Organic Chemistry

b.
```
    H H H H H H H H H H
    | | | | | | | | | |
H - C-C-C-C-C-C-C-C-C-C - H
    | | | | | | | | | |
    H H H H H H H H H H
```

c.
```
    H H H H H H H H H H H H
    | | | | | | | | | | | |
H - C-C-C-C-C-C-C-C-C-C-C-C - H
    | | | | | | | | | | | |
    H H H H H H H H H H H H
```

14. a.
```
    H H H H
    | | | |
H - C-C-C-C - H
    | | | |
    H H H H
```

butane $CH_3CH_2CH_2CH_3$

b.
```
    H H H H H H H H
    | | | | | | | |
H - C-C-C-C-C-C-C-C - H
    | | | | | | | |
    H H H H H H H H
```

octane $CH_3CH_2CH_2CH_2CH_2CH_2CH_2CH_3$

c.
```
    H H H H H
    | | | | |
H - C-C-C-C-C - H
    | | | | |
    H H H H H
```

pentane $CH_3CH_2CH_2CH_2CH_3$

d.
```
    H H H H H H
    | | | | | |
H - C-C-C-C-C-C - H
    | | | | | |
    H H H H H H
```

hexane $CH_3CH_2CH_2CH_2CH_2CH_3$

16. branch or substituent

18. With eight carbon atoms, there are a *lot* of isomers possible. Here are carbon skeletons for some of them:

```
C-C-C-C-C-C-C-C    C-C-C-C-C-C-C    C-C-C-C-C-C-C    C-C-C-C-C-C-C
                         |                   |                   |
                         C                   C                   C
```

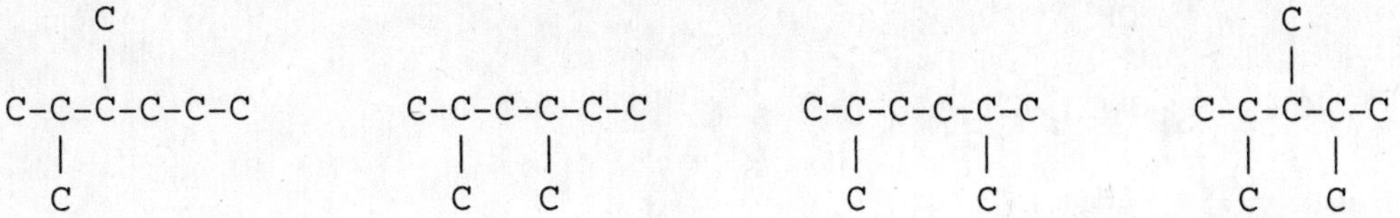

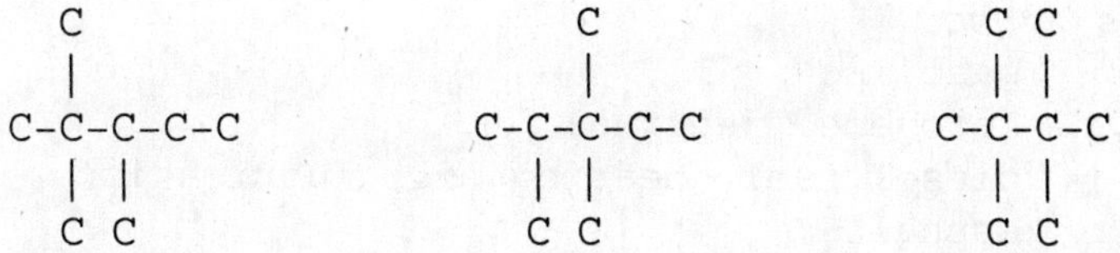

```
            C                                                              C
            |                                                              |
    C-C-C-C-C-C         C-C-C-C-C-C         C-C-C-C-C-C         C-C-C-C-C
            |               |   |               |   |               |   |
            C               C   C               C   C               C   C


        C                       C                   C C
        |                       |                   | |
    C-C-C-C-C             C-C-C-C-C             C-C-C-C       (plus 7 others)
      | |                   | |                   | |
      C C                   C C                   C C
```

20. The root name is derived from the number of carbon atoms in the *longest continuous chain* of carbon atoms.

22. The position of a substituent is indicated by a number which corresponds to the carbon atom in the longest chain to which the substituent is attached.

24. Multiple substituents are listed in alphabetical order, disregarding any prefix.

26. a. 3-ethylpentane

    b. 2,2-dimethylbutane

    c. 2,2-dimethylpropane

    d. 2,3,4-trimethylpentane

28. a.
```
    CH₃-CH-CH₂-CH₂-CH₂-CH₃
        |
        CH₃
```

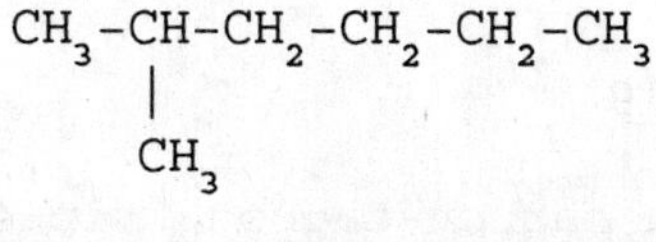

    b.
```
    CH₃-CH₂-CH-CH₂-CH₂-CH₃
            |
            CH₃
```

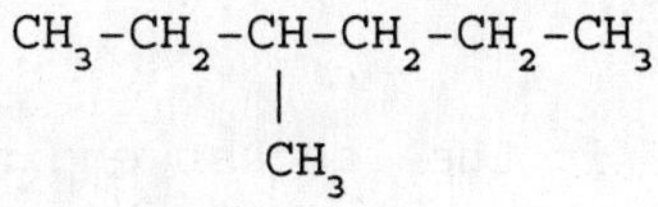

    c.
```
            CH₃
            |
    CH₃-C-CH₂-CH₂-CH₂-CH₃
            |
            CH₃
```

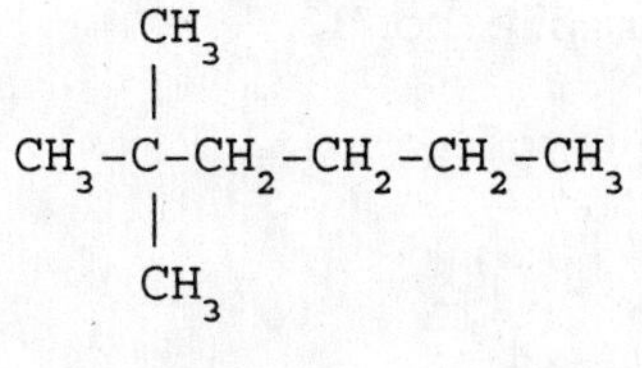

    d.
```
                CH₃
                |
    CH₃-CH-CH-CH₂-CH₂-CH₃
        |
        CH₃
```

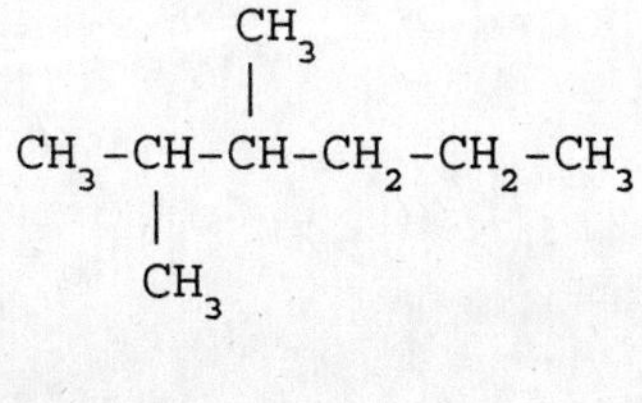

248   Chapter 19   Organic Chemistry

e.
$$CH_3-CH_2-\underset{\underset{CH_3}{|}}{\overset{\overset{CH_3}{|}}{C}}-CH_2-CH_2-CH_3$$

30.
Number of C atoms	Use
C_5-C_{12}	gasoline
$C_{10}-C_{18}$	kerosene, jet fuel
$C_{15}-C_{25}$	diesel fuel, heating oil, lubrication
$C_{25}-$	asphalt

32. Tetraethyl lead was added to gasolines to prevent "knocking" of high efficiency automobile engines. The use of tetraethyl lead is being phased out because of the danger to the environment of the lead in this substance.

34. Combustion represents the vigorous reaction of a hydrocarbon (or other substance) with oxygen. The combustion of alkanes has been made use of as a source of heat, light, and mechanical energy.

36. Dehydrogenation reactions involve the removal of hydrogen atoms from adjacent carbon atoms in an alkane (or other substance). When two hydrogen atoms are removed from an alkane or related compound, a double bond is created.

38. a. $2C_6H_{14}(l) + 19O_2(g) \rightarrow 12CO_2(g) + 14H_2O(g)$

b. $CH_4(g) + Cl_2(g) \rightarrow CH_3Cl(l) + HCl(g)$

c. $CHCl_3(l) + Cl_2(g) \rightarrow CCl_4(l) + HCl(g)$

40. An alkyne is a hydrocarbon containing a carbon-carbon triple bond. The general formula is C_nH_{2n-2}.

42. The location of a double or triple bond in the longest chain of an alkene or alkyne is indicated by giving the *number* of the lowest number carbon atom involved in the double or triple bond.

44. hydrogenation

46. a. 1-butyne

b. 3-methyl-1-butyne

c. 3-heptyne

48. Shown are carbon skeletons:

   C≡C-C-C-C-C   C-C≡C-C-C-C   C-C-C≡C-C-C   C≡C-C-C-C
                                                      | |
                                                      C   C

   C-C≡C-C-C   C≡C-C-C-C   C≡C-C-C
      | | |
      C   C   C

50. For benzene, a *set* of equivalent Lewis structures can be drawn, differing only in the *location* of the three double bonds in the ring. Experimentally, however, benzene does not demonstrate the chemical properties expected for molecules having *any* double bonds. We say that the "extra" electrons that would go into making the second bond of the three double bonds are delocalized around the entire benzene ring: this delocalization of the electrons explains benzene's unique properties.

52. When named as a substituent, the benzene ring is called the *phenyl* group. Two examples are

    CH$_2$=CH—CH—CH$_3$ CH$_3$—CH—CH$_2$—CH$_2$—CH$_2$—CH$_3$
           | |
        (phenyl)   (phenyl)

    3-phenyl-1-butene   2-phenylhexane

54. *ortho-* refers to adjacent substituents (1,2-); *meta-* refers to two substituents with one unsubstituted carbon atom between them (1,3-); *para-* refers to two substituents with two unsubstituted carbon atoms between them (1,4-).

56. a.   1,2-dimethylbenzene (*o*-xylene is its common name)
    b.   1,2,3,4,5,6-hexachlorobenzene
    c.   anthracene
    d.   3,5-dichloro-1-methylbenzene (3,5-dichlorotoluene is another name)

58. a.   ether   b.   alcohol
    c.   alcohol   d.   organic (carboxylic) acid

60. Primary alcohols have *one* hydrocarbon fragment (alkyl group) bonded to the carbon atom where the –OH group is attached. Secondary alcohols have *two* such alkyl groups attached, and tertiary alcohols contain *three* such alkyl groups. Examples are

    ethanol (primary)   CH$_3$-CH$_2$-OH

2-propanol (secondary) $CH_3-CH-CH_3$
                                      |
                                     OH

                                     CH_3
                                      |
2-methyl-2-propanol (tertiary) CH_3-C-CH_3
                                      |
                                     OH

62. a. $CH_3-CH_2-CH_2-CH_2-CH_2-OH$ primary

    b. $CH_3-CH-CH_2-CH_2-CH_3$ secondary
             |
            OH

    c. $CH_3-CH_2-CH-CH_2-CH_3$ secondary
                  |
                 OH

            OH
             |
    d. $CH_2-CH_2-C-CH_2-CH_3$ tertiary
                  |
                CH_3

64. $C_6H_{12}O_6$ –yeast→ $2CH_3-CH_2-OH + 2CO_2$

    The yeast necessary for the fermentation process are killed if the concentration of ethanol is over 13%. More concentrated ethanol solutions are most commonly made by distillation.

66. methanol (CH_3OH) - starting material for synthesis of acetic acid and many plastics

    ethylene glycol (CH_2OH-CH_2OH) - automobile antifreeze

    isopropyl alcohol (2-propanol, $CH_3-CH(OH)-CH_3$) - rubbing alcohol

68. Aldehydes and ketones both contain the carbonyl group $\diagdown \hspace{-0.7em} \diagup \hspace{-0.5em} C=O$.

Aldehydes and ketones differ in the *location* of the carbonyl function: aldehydes contain the carbonyl group at the end of a hydrocarbon chain (the carbon atom of the carbonyl group is bonded only to at most one other carbon atom); the carbonyl group of ketones represents one of the interior carbon atoms of a chain (the carbon atom of the carbonyl group is bonded to two other carbon atoms).

70. Aldehydes and ketones are produced by the oxidation of primary and secondary alcohols, respectively.

$$CH_3-CH_2-OH \xrightarrow{\text{oxidation}} CH_3-\underset{H}{\overset{\displaystyle C=O}{|}}$$

$$CH_3-\underset{CH_3}{\overset{\displaystyle CH-OH}{|}} \xrightarrow{\text{oxidation}} CH_3-\underset{CH_3}{\overset{\displaystyle C=O}{|}}$$

72. In addition to their systematic names (based on the hydrocarbon root, with the ending *-one*), ketones can also be named by naming the groups attached to either side of the carbonyl carbon as alkyl groups, followed by the word "ketone". Examples are

$CH_3-C(=O)-CH_2CH_3$ methyl ethyl ketone (2-butanone, butanone)

$CH_3CH_2-C(=O)-CH_2CH_3$ diethyl ketone (3-pentanone)

74. a. $CH_3-\underset{\displaystyle O}{\overset{\displaystyle \|}{C}}-\phi$ [ϕ represents the phenyl group (benzene ring)]

   b. $CH_3-CH_2-CH_2-\underset{H}{\overset{\displaystyle C=O}{|}}$

   c. $CH_3-CH_2-\underset{\displaystyle O}{\overset{\displaystyle \|}{C}}-CH_3$

   d. $CH_3-CH_2-CH_2-\underset{\displaystyle O}{\overset{\displaystyle \|}{C}}-CH_2-CH_2-CH_3$

   e. $CH_3-CH_2-CH_2-\underset{\displaystyle O}{\overset{\displaystyle \|}{C}}-CH_2-CH_2-CH_3$ (same molecule as above)

76. Carboxylic acids are typically *weak* acids.

$CH_3-CH_2-COOH(aq) \rightleftarrows H^+(aq) + CH_3-CH_2-COO^-(aq)$

78. a. CH_2-CH_2-COOH

   b. $CH_3-CH_2-\underset{\underset{O}{\|}}{C}-CH_3$

   c. no reaction (tertiary alcohols are not oxidized in this manner)

80. Acetylsalicylic acid is synthesized from salicylic acid (behaving as an alcohol through its -OH group) and acetic acid.

82. a. $CH_3-CH_2-\underset{\underset{CH_3}{|}}{CH}-\underset{\underset{OH}{|}}{CH_2}-C=O$

   b. $H-\underset{\underset{O}{\|}}{C}-O-CH_2-CH_3$

   c. $\phi-\underset{\underset{O}{\|}}{C}-O-CH_3$ [ϕ represents the pheynyl group (benzene ring)]

   d. $CH_3-CH_2-\underset{\underset{Br}{|}}{CH}-\underset{\underset{OH}{|}}{C}=O$

   e. $CH_3-CH_2-\underset{\underset{CH_3}{|}}{\overset{\overset{Cl}{|}}{CH}}-\overset{\overset{OH}{|}}{CH}-\underset{\underset{CH_3}{|}}{CH}-C=O$

84. In addition polymerization, the monomer units simply add together to form the polymer, with no other products. Polyethylene and polytetrafluoroethylene (Teflon) are common examples.

86. A polyester is formed from the reaction of a dialcohol (two -OH groups) with a diacid (two -COOH groups). One -OH group of the alcohol forms an *ester linkage* with one of the -COOH groups of the acid. Since the resulting dimer still possesses an -OH and a -COOH group, the dimer can undergo further esterification reactions. Dacron is a common polyester.

88. **nylon**

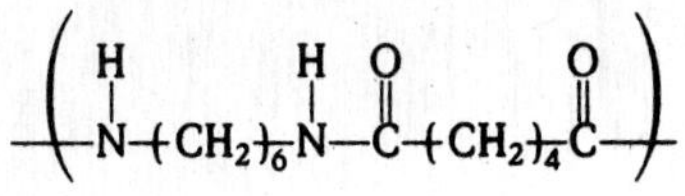

$$-\left(\begin{array}{c}H\\|\\N\end{array}-(CH_2)_6-\begin{array}{c}H\\|\\N\end{array}-\begin{array}{c}O\\||\\C\end{array}-(CH_2)_4-\begin{array}{c}O\\||\\C\end{array}-\right)-$$

dacron

$$-\left(O-CH_2-CH_2-O-\overset{O}{\underset{||}{C}}-\underset{}{\bigcirc}-\overset{O}{\underset{||}{C}}\right)-$$

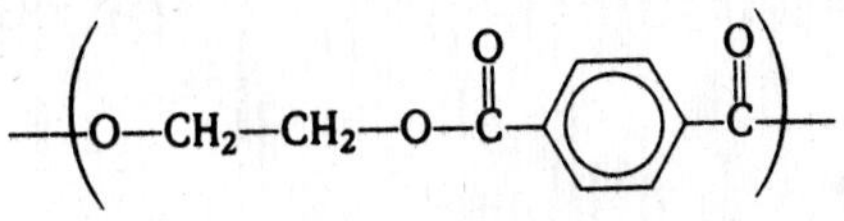

90. saturated

92. straight-chain or normal

94. -ane

96. number

98. anti-knocking

100. substitution

102. hydrogenation

104. functional

106. carbon monoxide

108. carbonyl

110. carboxyl

112. addition

114. The carbon skeletons are

```
                                          C
                                          |
  C-C-C-C-C         C-C-C-C         C-C-C
                        |               |
                        C               C
```

116. a. 2-chlorobutane
      b. 1,2-dibromoethane
      c. triiodomethane (common name: iodoform)
      d. 2,3,4-trichloropentane
      e. 2,2-dichloro-4-isopropylheptane

118. a.
```
                CH3
                |
      CH3-CH-CH-CH2-CH2-CH2-CH3
           |
           CH3
```

b. HO-CH$_2$-C(CH$_3$)(CH$_3$)-CH(Cl)-CH$_2$-CH$_2$-CH$_2$-CH$_2$-CH$_3$

c. CH$_2$=C(Cl)-CH$_2$-CH$_2$-CH$_2$-CH$_3$

d. CH$_2$(Cl)-CH=CH-CH$_2$-CH$_2$-CH$_3$

e. 2-methylphenol (o-cresol): benzene ring with OH and CH$_3$ on adjacent carbons

120. primary   CH$_3$-CH$_2$-CH$_2$-CH$_2$-CH$_2$-CH$_2$-OH

  secondary CH$_3$-CH$_2$-CH$_2$-CH$_2$-CH(OH)-CH$_3$

  tertiary   CH$_3$-CH$_2$-C(CH$_3$)(OH)-CH$_2$-CH$_3$

122. CH$_2$(OH)-CH(OH)-CH(OH)-CH(OH)-CH(OH)-C(=O)H

124. a. CH$_3$-C(=O)-CH$_2$-CH$_2$-CH$_2$-CH$_2$-CH$_3$

  b. CH$_3$-CH$_2$-CH(CH$_3$)-CH$_2$-C(=O)H

  c. CH$_3$-CH$_2$-CH$_2$-CH(CH$_3$)-CH$_2$-OH

  d. CH$_2$(OH)-CH(OH)-CH$_2$(OH)

e. 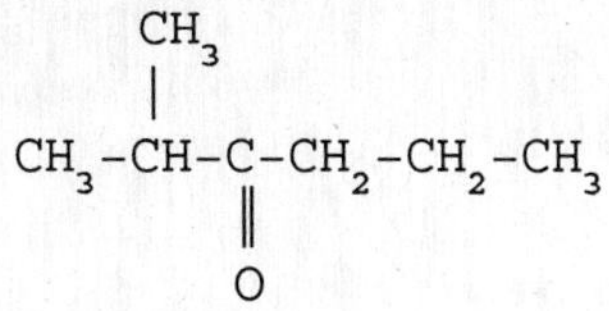
$$CH_3-CH-\underset{\underset{O}{\|}}{\underset{|}{C}}-CH_2-CH_2-CH_3$$
$$\overset{|}{CH_3}$$

126.

$$\underset{\underset{H}{|}}{\overset{\overset{HO-C=O}{|}}{CH_3-CH-N}}\!\!-\!\!\overparen{(H+HO)}\!\!-\!\!\underset{CH_2-NH_2}{C=O} \rightarrow$$

$$\underset{\underset{H}{|}}{\overset{\overset{HO-C=O}{|}}{CH_3-CH-N}}\!\!-\!\!\underset{CH_2-NH_2}{\overset{|}{C=O}} + H_2O$$

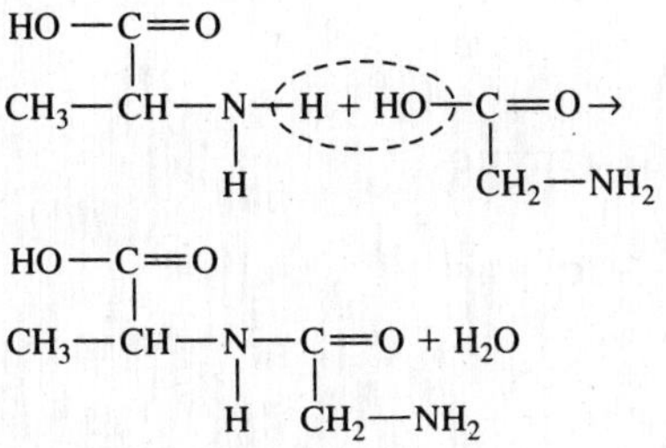

One end has —NH_2, which can react with the —COOH end of another of these dipeptides.

128. a.

```
   H H H H H H H H
   | | | | | | | |
H- C-C-C-C-C-C-C-C -H
   | | | | | | | |
   H H H H H H H H
```

octane $CH_3CH_2CH_2CH_2CH_2CH_2CH_2CH_3$

b.
```
   H H H H H H
   | | | | | |
H- C-C-C-C-C-C -H
   | | | | | |
   H H H H H H
```

hexane $CH_3CH_2CH_2CH_2CH_2CH_3$

c.
```
   H H H H
   | | | |
H- C-C-C-C -H
   | | | |
   H H H H
```

butane $CH_3CH_2CH_2CH_3$

d.
```
   H H H H H
   | | | | |
H- C-C-C-C-C -H
   | | | | |
   H H H H H
```

pentane $CH_3CH_2CH_2CH_2CH_3$

130. a. 2,3-dimethylbutane
b. 3,3-diethylpentane
c. 2,3,3-trimethylhexane
d. 2,3,4,5,6-pentamethylheptane

256   Chapter 19   Organic Chemistry

132. a. $CH_3Cl(g)$
     b. $H_2(g)$
     c. $HCl(g)$

134. $CH\equiv C-CH_2-CH_2-CH_2-CH_2-CH_2-CH_3$ 1-octyne

     $CH_3-C\equiv C-CH_2-CH_2-CH_2-CH_2-CH_3$ 2-octyne

     $CH_3-CH_2-C\equiv C-CH_2-CH_2-CH_2-CH_3$ 3-octyne

     $CH_3-CH_2-CH_2-C\equiv C-CH_2-CH_2-CH_3$ 4-octyne

136. a. carboxylic acid
     b. ketone
     c. ester
     d. alcohol (phenol)

138. a. 3-methylpentanal

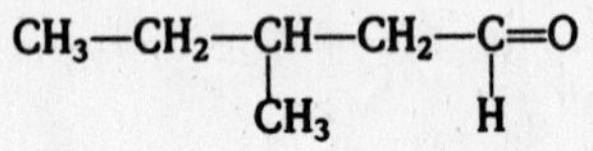

$$CH_3-CH_2-\underset{\underset{CH_3}{|}}{CH}-CH_2-\underset{\underset{H}{|}}{C}=O$$

   b. 3-methyl-2-pentanone

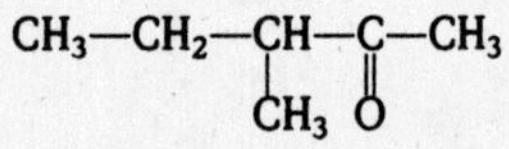

$$CH_3-CH_2-\underset{\underset{CH_3}{|}}{CH}-\underset{\underset{O}{\|}}{C}-CH_3$$

   c. methyl phenyl ketone

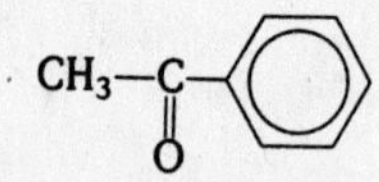

$$CH_3-\underset{\underset{O}{\|}}{C}-C_6H_5$$

   d. 2-hydroxybutanal

$$CH_3-CH_2-\underset{\underset{OH}{|}}{CH}-\underset{\underset{H}{|}}{C}=O$$

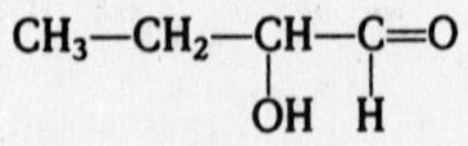

   e. propanal

$$CH_3-CH_2-\underset{\underset{H}{|}}{C}=O$$

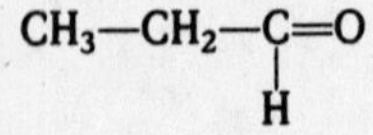

140. a. CH$_3$-CH-CH$_2$-COOH
           |
           CH$_3$

b.

[benzene ring with -C(=O)-OH group and -Cl substituent ortho to it]

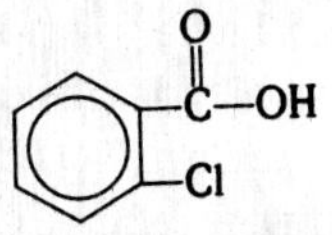

c. CH$_3$-CH$_2$-CH$_2$-CH$_2$-CH$_2$-COOH

d. CH$_3$-COOH

Chapter 20   Biochemistry

2. Proteins represent biopolymers of α-amino acids used for many purposes in the human body (structure, enzymes, antibodies, etc.). Proteins make up about 15% of the body by mass.

4. Fibrous proteins provide the structural material of many tissues in the body, and are the chief constituents of hair, cartilage, and muscles. Fibrous proteins consist of lengthwise bundles of polypeptide chains (a fiber). Globular proteins have their polypeptide chains folded into a basically spherical shape and tend to be found in the bloodstream, where they transport and store various needed substances, act as antibodies to fight infections, act as enzymes to catalyze cellular processes, participate in the body's various regulatory systems, and so on.

6. The structures of the amino acids are given in detail in Figure 20.2. Generally a side chain is nonpolar if it is mostly hydrocarbon in nature (e.g., alanine, in which the side chain is a methyl group). Side chains are polar if they contain the hydroxyl group (-OH), the sulfhydryl group (-SH), or a second amino (-NH$_2$) or carboxyl (-COOH) group.

8. Figure 20.2 shows the amino acids separated into the hydrophilic and hydrophobic subgroupings. Notice that the hydrophobic amino acids contain R groups which are basically hydrocarbon (or substituted hydrocarbon) in nature. This makes these R groups nonpolar and unlikely to interact with very polar water molecules. Notice that the hydrophilic amino acids all contain a very polar functional group (e.g., -OH, -SH, -C=O, etc.) which enable the amino acid R groups to interact with water. For a protein in an aqueous medium, the hydrophilic R groups will orient themselves towards the aqueous medium, while the hydrophobic R groups will turn away from the aqueous medium, causing a profound effect on the protein's shape.

10. There are six tripeptides possible.

    cys-ala-phe   ala-cys-phe   phe-ala-cys

    cys-phe-ala   ala-phe-cys   phe-cys-ala

12. phe-ala-gly

    phe-gly-ala

ala-phe-gly

$$H_2N-\underset{\underset{N\ terminal}{}}{\overset{CH_3}{\underset{|}{C}}}-\overset{O}{\underset{}{C}}-N-\overset{CH_2}{\underset{|}{C}}-\overset{O}{\underset{}{C}}-N-\overset{H}{\underset{|}{C}}-COOH$$
(with phenyl ring on CH₂ of middle residue; peptide bonds circled)

ala-gly-phe

$$H_2N-\overset{CH_3}{\underset{|}{C}}-\overset{O}{C}-N-\overset{H}{\underset{|}{C}}-\overset{O}{C}-N-\overset{CH_2}{\underset{|}{C}}-COOH$$
(phenyl ring on C-terminal CH₂)

gly-phe-ala

$$H_2N-\overset{H}{\underset{|}{C}}-\overset{O}{C}-N-\overset{CH_2}{\underset{|}{C}}-\overset{O}{C}-N-\overset{CH_3}{\underset{|}{C}}-COOH$$
(phenyl ring on middle CH₂)

gly-ala-phe

$$H_2N-\overset{H}{\underset{|}{C}}-\overset{O}{C}-N-\overset{H}{\underset{|}{C}}-\overset{O}{C}-N-\overset{CH_2}{\underset{|}{C}}-COOH$$
(phenyl ring on C-terminal CH₂)

14. Long, thin, resilient proteins, such as hair, typically contain elongated, elastic alpha-helical protein molecules. Other proteins, such as silk, which in bulk form sheets or plates, typically contain protein molecules having the beta pleated sheet secondary structure. Proteins which do not have a structural function in the body, such as hemoglobin, typically have a globular structure.

16. The secondary structure, in general, describes the arrangement of the long polypeptide chain of the protein. In the alpha-helical secondary structure, the chain forms a coil or spiral, which gives proteins consisting of such structures an elasticity or resilience. Such proteins are found in wool, hair, and tendons.

18. Cysteine, an amino acid containing the sulfhydryl (-SH) group in its side chain, is capable of forming disulfide linkages (-S-S-) with other cysteine molecules in the same polypeptide chain. If such a disulfide

linkage is formed, this effectively ties together two portions of the polypeptide, producing a kink or knot in the chain, which leads in part to the protein's overall 3-dimensional shape (tertiary structure). Cysteine, and the disulfide linkages it forms, is responsible for the curling of hair (whether naturally or by a permanent wave).

20. Collagen has an alpha-helical secondary structure. Collagen's function in the body is as the raw material of which tendons are constructed. The long, springy structure of the alpha-helix is responsible for collagen's strength and elasticity.

22. Antibodies are special proteins which are synthesized in the body in response to foreign substances such as bacteria or viruses. Although there are usually specific antibodies for specific invaders, the antibody interferon offers general protection against invasion by viruses.

24. In a permanent wave, cross-linkages between adjacent polypeptide chains of the protein are broken chemically, and then reformed chemically in a new location. The primary cross-linkage involved is a disulfide linkage between cysteine units in the polypeptide chains. It is primarily the tertiary structure of the hair protein which is affected by a permanent wave, although if the waving lotion is left on the hair too long, the secondary structure can also be affected (making the hair very "frizzy").

26. The molecule acted upon by an enzyme is referred to as the enzyme's substrate. When we say that an enzyme is specific for a particular substrate, we mean that the enzyme will catalyze the reactions of that molecule and that molecule only.

28. Figure 20.11 illustrates the lock and key model clearly. The lock-and-key model for enzyme action postulates that the structures of the enzyme and its substrate are complementary, such that the active site of the enzyme and the portion of the substrate molecule to be acted upon can fit together very closely. The structures of these portions of the molecules are unique to the particular enzyme-substrate pair, and they fit together much like a particular key is necessary to work in a given lock.

30. Sugars contain an aldehyde or ketone functional group (carbonyl group), as well as several (OH groups (hydroxyl group).

32. a. glucose
```
         CHO
          |
     H — C — OH
          |
    HO — C — H
          |
     H — C — OH
          |
     H — C — OH
          |
         CH₂OH
```

b. ribose

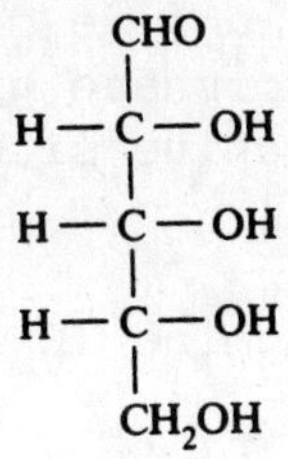

$$\begin{array}{c} \text{CHO} \\ \text{H}-\text{C}-\text{OH} \\ \text{H}-\text{C}-\text{OH} \\ \text{H}-\text{C}-\text{OH} \\ \text{CH}_2\text{OH} \end{array}$$

c. ribulose

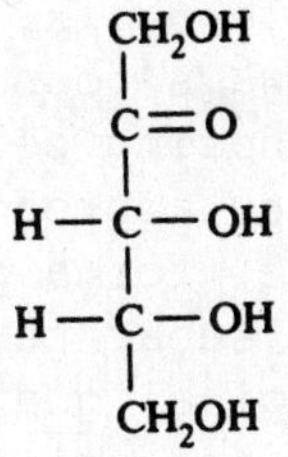

$$\begin{array}{c} \text{CH}_2\text{OH} \\ \text{C}=\text{O} \\ \text{H}-\text{C}-\text{OH} \\ \text{H}-\text{C}-\text{OH} \\ \text{CH}_2\text{OH} \end{array}$$

d. galactose

$$\begin{array}{c} \text{CHO} \\ \text{H}-\text{C}-\text{OH} \\ \text{HO}-\text{C}-\text{H} \\ \text{HO}-\text{C}-\text{H} \\ \text{H}-\text{C}-\text{OH} \\ \text{CH}_2\text{OH} \end{array}$$

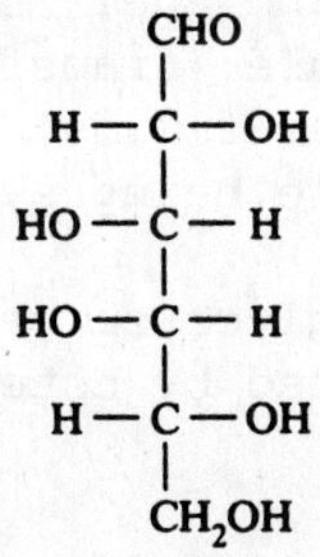

34. A polysaccharide is a large polymeric substance, containing as its building blocks repeating simple sugar (monosaccharide) monomer units. Both starch and cellulose consist of long chains of bonded *glucose* molecules: the differences in properties between starch and cellulose are derived from the different manner in which the glucose units are attached to one another to form the polysaccharide chains.

36. ribose (aldopentose); arabinose (aldopentose); ribulose (ketopentose); glucose (aldohexose); mannose (aldohexose); galactose (aldohexose); fructose (ketohexose).

38. A nucleotide consists of three components: a five-carbon sugar, a nitrogen-containing organic base, and a phosphate group. The phosphate and nitrogen base are each bonded to respective sites on the sugar, but are not bonded to each other.

40. Uracil (RNA only); cytosine (DNA, RNA); thymine (DNA only); adenine (DNA, RNA); guanine (DNA, RNA)

42. When the two strands of a DNA molecule are compared, it is found that a given base in one strand is always found paired with a particular base in the other strand. Because of the shapes and side atoms along the rings of the nitrogen bases, only certain pairs are able to approach and hydrogen-bond with each other in the double helix. Adenine is always found paired with thymine; cytosine is always found paired with guanine. When a DNA helix unwinds for replication during cell division, only the appropriate complementary bases are able to approach and bond to the nitrogen bases of each strand. For example, for a guanine-cytosine pair in the original DNA, when the two strands separate, only a new cytosine molecule can approach and bond to the original guanine, and only a new guanine molecule can approach and bond to the original cytosine.

44. Messenger RNA molecules are synthesized to be complementary to a portion (gene) of the DNA molecule in the cell, and serve as the template or pattern upon which a protein will be constructed (a particular group of nitrogen bases on *m*-RNA is able to accommodate and specify a particular amino acid in a particular location in the protein). Transfer RNA molecules are much smaller than *m*-RNA, and their structure accommodates only a single specific amino acid molecule: transfer RNA molecules "find" their specific amino acid in the cellular fluids, and bring it to *m*-RNA where it is added to the protein molecule being synthesized.

46. A triglyceride typically consists of a glycerol backbone, to which three separate fatty acid molecules are attached by ester linkages

$$\begin{array}{l} CH_2-O-\overset{\overset{O}{\|}}{C}-R \\ CH-O-\overset{\overset{O}{\|}}{C}-R' \\ CH_2-O-\overset{\overset{O}{\|}}{C}-R'' \end{array}$$

48. Saponification is the production of a *soap* by treatment of a triglyceride with a strong base such as NaOH.

   triglyceride + 3NaOH → glycerol + 3Na⁺soap⁻

$$\begin{array}{l} CH_2-O-\overset{\overset{O}{\|}}{C}-R \\ CH-O-\overset{\overset{O}{\|}}{C}-R' \\ CH_2-O-\overset{\overset{O}{\|}}{C}-R'' \end{array} + 3NaOH \rightarrow \begin{array}{l} CH_2-OH \\ CH-OH \\ CH_2-OH \end{array} + \begin{array}{l} RCOONa \\ R'COONa \\ R''COONa \end{array}$$

50. Soaps have both a nonpolar nature (due to the long chain of the fatty acid) and an ionic nature (due to the charge on the carboxyl group). In water, soap anions form aggregates called micelles, in which the water-repelling hydrocarbon chains are oriented towards the interior of the aggregate, with the ionic, water-attracting carboxyl groups oriented towards the outside. Most dirt has a greasy nature. A soap micelle is able to interact with a grease molecule, pulling the grease molecule into the hydrocarbon interior of the micelle. When the clothing is rinsed, the micelle containing the grease is washed away. See Figures 20.22 and 20.23.

52. Cholesterol is the naturally occurring steroid from which the body synthesizes other needed steroids. Since cholesterol is insoluble in water, it is thought that having too large a concentration of this substance in the bloodstream may lead to its deposition and build up on the walls of blood vessels, causing their eventual blockage.

54. Phospholipids have a similar structure to triglycerides, except that one of the fatty acids attached to the glycerol backbone is replaced by a phosphate group (phospholipids are sometimes referred to as phosphodiglycerides). An important phospholipid is lecithin, which behaves very strongly as an emulsifying agent (allowing polar and nonpolar substances to mix). Lecithin is found in high concentration in egg yolks, and is being used in baking in place of whole eggs: the emulsifying properties of the egg yolk are due to the lecithin, and using lecithin itself avoids the cholesterol problem associated with whole eggs.

56.	i	58.	m
60.	u	62.	f
64.	g	66.	r
68.	p	70.	o
72.	b	74.	d
76.	a	78.	nucleotides
80.	ester	82.	thymine, guanine
84.	transfer, messenger	86.	lipids
88.	unsaturated, saturated	90.	ionic, nonpolar
92.	fatty (long chain)	94.	progesterone

96. The primary structure of a protein refers to the specific identity and ordering of amino acids in a protein's polypeptide chain. The primary structure is sometimes referred to as the protein's amino acid *sequence*.

98. tendons, bone (with mineral constituents), skin, cartilage, hair, fingernails.

100. Proteins contain both acidic (-COOH) and basic ($-NH_2$) groups in their side chains, which can neutralize both acids and bases.

102. pentoses (5 carbons); hexoses (6 carbons); trioses (3 carbons)

104. In a strand of DNA, the phosphate group and the sugar molecule of adjacent nucleotides become bound to each other. The chain-portion of the DNA molecule, therefore, consists of alternating phosphate groups and sugar molecules. The nitrogen bases are found sticking out from the side of this phosphate-sugar chain, being bonded to the sugar molecules.

106. The primary human bile acid is cholic acid, which helps to emulsify fats in the intestinal tract. In order to be digested by enzymes for absorption into the bloodstream, large clumps of fat must be dispersed into fine droplets in the liquid of the small intestine. Cholic acid basically acts as a detergent.